AF784933

TWIN-SCREW EXTRUDERS: A BASIC UNDERSTANDING

HOECHST CELANESE T.I.C. CORPUS CHRISTI
1050011415

TWIN-SCREW EXTRUDERS: A BASIC UNDERSTANDING

Fabrizio G. Martelli, Ph.D.

HOECHST CELANESE
INFORMATION CENTER
1901 CLARKWOOD RD.
P. O. BOX 9077
CORPUS CHRISTI, TX 78469-9077

VAN NOSTRAND REINHOLD COMPANY
NEW YORK CINCINNATI TORONTO LONDON MELBOURNE

Copyright © 1983 by Van Nostrand Reinhold Company Inc.

Library of Congress Catalog Card Number: 81-19779
ISBN: 0-442-26363-5

All rights reserved. No part of this work covered by the copyright hereon may be reproduced or used in any form or by any means–graphic, electronic, or mechanical, including photocopying, recording, taping, or information storage and retrieval systems–without permission of the publisher.

Manufactured in the United States of America

Published by Van Nostrand Reinhold Company Inc.
135 West 50th Street, New York, N.Y. 10020

Van Nostrand Reinhold Publishing
1410 Birchmount Road
Scarborough, Ontario M1P 2E7, Canada

Van Nostrand Reinhold Australia Pty. Ltd.
17 Queen Street
Mitcham, Victoria 3132, Australia

Van Nostrand Reinhold Company Limited
Molly Millars Lane
Wokingham, Berkshire, England

15 14 13 12 11 10 9 8 7 6 5 4 3 2 1

Library of Congress Cataloging in Publication Data

Martelli, Fabrizio G.
Twin-screw extruders.

Includes index.
1. Plastics machinery. 2. Plastics–Extrusion.
I. Title.
TP1135.M37 668.4'13'028 81-19779
ISBN 0-442-26363-5 AACR2

Foreword

Most books on plastics machinery include a preamble on the origin of such equipment, and some even discuss the origin of plastic itself, dating back to the early 1900s and such men as Leo Baekeland – the real founder of synthetic plastics. There seems therefore, little purpose in reiterating what has been said before and going over the same ground so adequately covered in a number of books as well as the trade press. We are indebted to the author of this excellent treatise on twin-screw extruders for getting right down to the business at hand.

The author makes mention of two pioneers – Roberto Colombo and Carlo Pasquetti – who were the first to develop twin-screw extruders. It was my good fortune to follow the work of these pioneers, and, interestingly enough, the principles were so good that their work continues to be relevant even to the advanced and more sophisticated models so well defined in this book.

The author, having spent more than thirty years studying and operating these machines the world over, is obviously an expert in this field. Therefore, as a pioneer in the field of plastics myself (over fifty years in the industry) I welcome a treatise on twin-screw extruders. Not much has been written on this subject except for a few articles that have appeared mostly in the trade press. Dr. Martelli's book gives us a basic understanding of this type of extruder. Misconceptions and perhaps some prejudices may be swept aside, and we may now know some of the advantages of twin-screw extruders.

As one who has purchased both single- and twin-screw extruders, I not only welcome this authoritative book, but I consider this as a valuable text book and a most valuable contribution to our ever-increasing library on plastics.

Charles A. Breskin

Preface

Extrusion equipment in recent years has been greatly improved and now does an excellent job in most applications. Yet there are still extrusion problems for which better solutions are needed.

Many tests and studies have been made to explain and characterize the extruder's work and to try to overcome existing difficulties. Almost all of them are concerned with single-screw extruders.

This type of extrusion machine, though well capable of processing thermoplastics, has some peculiarities that are inherent in the basic design of the machine and its operation, that creates problems in the use of some materials that for various reasons, are difficult to extrude.

A great lack of information exists on another type of extruder: twin-screw machines. Whereas most people in the plastics industry are very familiar with single screw extrusion, not too many have knowledge, gained from personal experience, of the twin-screw extruders. These have been in use for a long time, but they are only now beginning to be appreciated.

Although some publications on extruders and extrusion talk about twin-screw extruders, they dedicate to this subject in most instances only a few paragraphs without analyzing it thoroughly. In others, misconceptions and generalizations sometimes lead to misunderstandings, and there is a need for a simple explanation of the construction and operation of these types of machines.

These misconceptions reached the extent that one of the most common objections to twin-screw extruders from people who are not familiar with them, is that twin-screw extruders yield a lower output than single-screw extruders of the same screw size, and some of them even expect from a twin-screw extruder twice the production than from a single-screw extruder of the same screw diameter. This comparison, as well as this expectation, is basically wrong. Nobody thinks in terms of comparing cars by the size of the wheels, but a sensible way may be to compare the car's speed in relation to horsepower.

Another objection is that, as these extruders are equipped with motors of much lower power than single-screw extruders, they must be able to extrude only much lower amounts of material. All that, of course, is also wrong and many studies have been published on twin-screw extruders attempting to clarify their operation.

Some of the studies made on twin-screw extruders, however, are related to a system of screws with a certain configuration and a certain shape of flights and channels, so that the results cannot be generalized. Some of these studies apply to devices like kneting discs or barrel valves assembled on a particular type of twin-screw extruder.

Very little literature is available explaining the differences between the various types; statements like positive pumping, positive displacement pump, figure-eight motion of the material within the barrel or, conversely, C-shaped closed chambers often occur without further explanation as to why the pumping is positive or when the flow is continuous around the two screws or when the material remains locked in closed chambers.

A comparative look at the manner in which extruders process the material may clarify the substantial difference between single-screw extruders on the one hand and different types of twin-screw extruders on the other. It will show that twin-screw extruders have to be taken into very serious consideration when difficult or delicate thermoplastics are to be extruded.

Experimental data, collected over many years, have shown that twin-screw extruders definitely have certain performance advantages due to the particular geometry of the screws of these machines.

Fabrizio G. Martelli

Contents

TWIN-SCREW EXTRUDERS: A BASIC UNDERSTANDING

1
Single-Screw Extruders

Reviewing briefly the way single-screw extruders operate, we see that, basically, the screw rotating inside the barrel is not able, per se, to push the material forward. As a matter of fact, if for some reason the material filling the channels of the screw sticks to the screw, the latter becomes a rotating cylinder with no forwarding action.

To be pushed forward, the material should not rotate with the screw or at least should rotate at a slower rate than the screw. This compares with a bolt being turned while the nut turns with it: it cannot be tightened. Only when the nut is held fast will it move forward when the bolt is rotated.

The only force that can keep the material from turning with the screw and therefore make it advance along the barrel is its drag or friction against the inside surface of the barrel. The more friction, the less rotation of the material with the screw; the less rotation, the more forward motion.

To increase the frictional surface, the length of the barrel has been increased steadily. The ratio of the barrel length to the screw diameter (L/D) is of great importance with single-screw extruders; a large L/D ratio means more friction, which, in turn, means more propulsive action under the same extruding conditions. The material, of course, never moves in the barrel along a straight line. It will always have a certain amount of rotation around the screw together with a translation along the barrel; one turn of the screw does not make the material move forward for the length of one pitch.

The well-known drag-flow equation for single-screw extruder

$$Q = \frac{\pi^2 D^2 nh \sin \phi \cos \phi}{2}$$

is based on the supposition that a section of material across the rectangular channels of the screw, with an area of

$$A = \pi h D \sin \phi$$

moves along the channel with a speed of $n\pi D \cos \phi$, which is the component of the rotational speed along the channel (Fig. 1-1). This equation does not take into consideration the thickness of the flight ($b \cos \phi$) that should be subtracted from the area. Of much more importance in the drag-flow formula is the factor $\frac{1}{2}$. This is a theoretical integration constant that presupposes that:

1. The material within the channels behaves as a Newtonian fluid.
2. The viscosity is the same in every point of the section.
3. The velocity of the layer of material nearest the barrel is $n\pi D \cos \phi$.
4. The velocity of the layer of material nearest the screw is zero.

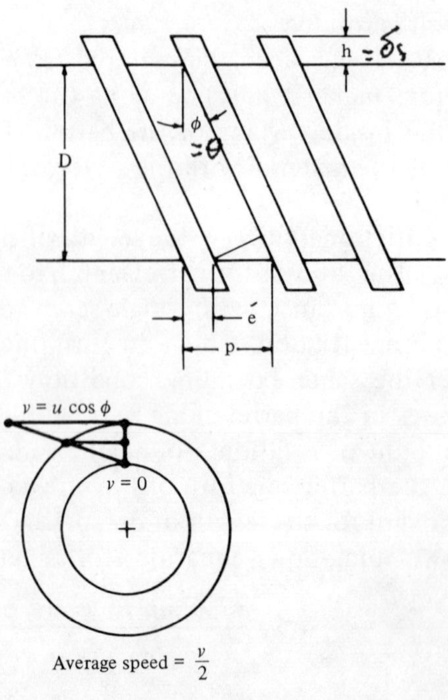

Fig. 1-1. Single screw extruder: geometry of the screw.

However, the viscosity is not the same at every point because most of the shear occurs near the periphery of the screw and not deep inside the channels. Furthermore, the velocity along the channel of the layer of material nearest to the screw cannot be zero. If the material has no velocity along the channel, it would, of course, not advance along the barrel, and this layer of polymer, however thin, would stay indefinitely inside the machine and would decompose or degrade, as the case may be.

The material, therefore, has to slide on the metal surfaces. This sliding depends on the frictional coefficients of the material on the inner surface of the barrel and of the screw.

These coefficients have been measured by Shooter and Thomas, who have given values in the speed range between 0.2″ and 24″ per minute (See Table 1-1).

Variations of the frictional coefficients with temperature have been measured by Gregory and others (Fig. 1-2). These frictional coefficients, as well as the viscosity, can be collected in a single factor F; more realistically, the velocity along the channel becomes $F(\pi D n \cos \phi)$. The drag-flow formula of the single-screw extruder is therefore better expressed by

$$Q = F(\pi^2 D^2 nh \sin \phi \cos \phi).$$

In perfect operating condition, this factor F may equal $\frac{1}{2}$, but if for example, the friction against the barrel is low and the friction against the screw is large and the viscosity is low, this factor can approach zero. Therefore, the drag flow may diminish from its maximum value to zero.

The machine output Q, which, when leakage flow is neglected, is equal to drag flow minus pressure flow, may become zero when F is equal to zero.

Table 1-1. Frictional Coefficients of Plastics.

Material	Frictional coefficients
Teflon	0.04
Polyethylene	0.15
Polystyrene	0.30
Acrylic	0.50

From Shooter and Thomas

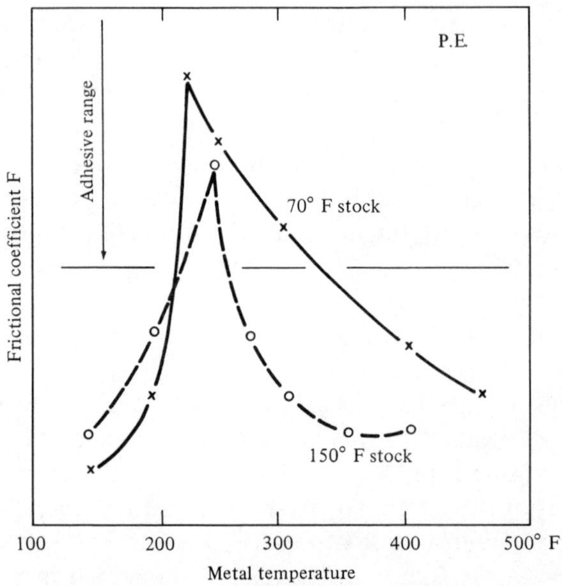

Fig. 1-2. Variation of the frictional coefficient with temperature (Gregory).

To yield enough production even with a low F factor, the screw has to turn at high speed and have a large diameter. However, a large-diameter screw rotating at high speed develops very high shear, as it appears from the shear rate equation

$$\dot{\gamma} = \frac{\pi D n}{h}$$

where D = screw diameter, n = screw speed, and h = channel depth. In non-Newtonian fluids, the viscosity is a function of shear as well as temperature. Higher screw speed means higher temperature and lower viscosity of the material; but lower viscosity, in turn, means smaller F factor and therefore lower output. Furthermore, higher temperature changes the frictional coefficients, again affecting F.

Friction of the material against the barrel, on which the entire functioning of a single-screw extruder depends, not only varies very much with the temperature but also is different for every material. As shown by Jacobi in his *Grundlagen der Extrudertechnik* (Fig. 1-3), the coefficient of friction of low-density polyethylene in a certain temperature range varies greatly with only a slight variation in temperature. This means that when this material is at a certain temper-

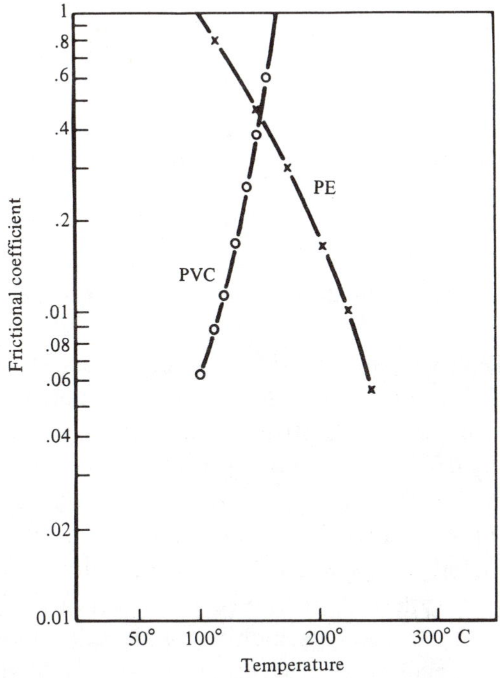

Fig. 1-3. Frictional coefficient of PVC and PE as function of the temperature (Jacobi).

ature, the friction is high, and therefore the temperature increases rapidly; once the temperature is high, the friction is low, and production of additional heat is also low. Rigid polyvinyl chloride (PVC), within the same temperature range, behaves in just the opposite way, and this characteristic makes it difficult to process it in an extruder that relies on friction for propulsion. When the material is cold, the frictional coefficient of the PVC is low, and therefore small heat production occurs; once the temperature has increased, the frictional coefficient increases greatly, and the temperature therefore increases rapidly.

If a certain friction is needed to propel the material through the extruder, the limit temperature for that material is easily exceeded, and degradation of the heat-sensitive polymer occurs. A decrease in shear rate, obtained by decreasing the screw speed, lowers the drag flow with the result that the extruder production decreases.

Another way to keep the shear rate low is by reducing the screw diameter, but this means again a decrease in production unless the screw speed is proportionally increased. Low shear rate can also be obtained by increasing the channel depth, but this affects the pressure flow by the cube of its value, therefore reducing production unless screw speed is again proportionally increased. To partially avoid this problem, the barrel has sometimes been provided with cooling devices, which, removing the excess frictional heat as soon as it is produced, keeps temperature down and viscosity high and creates friction of the material against the barrel. This increases the F factor, so that the same output can be obtained with a smaller-diameter screw, and lower speed.

Different materials and/or different dies may require different screw design. Other problems, not considered here, are inherent in the design of a single-screw extruder. Mixing, for example, is caused by the pressure difference between the front and rear of each flight and the different rotational velocity of the various layers, which forces the material to revolve inside the channel. This type of motion can mix particles adjacent to one another but not mix particles a few flights apart unless there is a large backflow at the land or tip of the screw flights. Surging, which is mainly due to variation of the coefficient F (especially in the melting zone) in self-sustaining oscillations, is another vexing problem in single-screw extruders.

2
Twin-Screw Extruders

The basic principle of twin-screw extruders for thermoplastics was conceived in Italy in the late thirties by Roberto Colombo of LMP (Lavorazione Materie Plastiche, Torino, Italy) while he was working with Carlo Pasquetti, also of Torino, on the problem of mixing cellulose acetate without the use of a solvent. Colombo developed a system of intermeshing corotating screws, which proved effective for the purpose; the result was, in fact, not only a mixer but also an extruder. Already, in 1939, a group of these machines were sold to the German I. G. Farbenindustrie for extrusion (Fig. 2-1).

Colombo's first patent on corotating twin-screw extruders was obtained in Italy (I. pat. 370578) on February 6, 1939; later, on August 7, 1947, after the interval caused by the war, this principle was patented by him in the United States (U.S. patent 2536396), then in Canada (C. patent 517911), England (B. patent 629109), and in other countries.

This new concept was found immediately so interesting that various companies acquired the rights to use these patents. Among the companies that manufactured these machines under the LMP/Colombo licence, for the length of the basic patent validity, were Windsor in England, Chemica in Switzerland, Herbosch-Polva in Belgium, Ikegai in Japan, and the CAFL, Creusot-Loire, in France.

Pasquetti, a little later, designed his own extruders adopting and patenting (B. pat. 677945) the counterrotating twin-screw extruders, and Schloemann was the first to use the Pasquetti system. Since then, twin-screw extruders have been developed to a high degree, and many different types of screws were made, each manufacturer finding his own solution for the various requirements of the extrusion process.

Some manufacturers, especially in the beginning, designed the screws uniformly all along, while others, keeping the flights similar to

Fig. 2-1. The very first corotating twin-screw extruder for thermoplastics (LMP/Colombo, 1939).

those of single-screw extruders, made shallow rectangular channels or channels shaped without regard to the configuration of the flights.

A better understanding of the twin-screw concept and of the importance of the shape of the flights – their depth, their intermeshing, and their perfect fit within the channels – is necessary to discriminate among these solutions.

Under the general name of twin-screw extruders are usually grouped together machines that, when thoroughly examined, turn out to have widely different characteristics. Among these extruders, we can distinguish various basic categories. Leaving aside extruders with multiple screws such as epicycloidal screws, extruders with one, two, or more screws added to a main screw, and other configurations that are mostly experimental, we will refer to extruders with only two screws of equal length placed inside the same barrel for their total length.

When talking about twin-screw extruders, it must be understood that this nomenclature should not apply to all extrusion machinery with two screws. Only too often, all extruders with two screws are wrongly grouped into the same category. The truth is that construction differences such as screw placement, shape of the flights, and direction of rotation make extruders as different from each other as they are, as a group, different from single-screw extruders.

The first who categorized the various types of twin-screw extruders was Erdmenger; he divided them according to the possible path of the material within the screws along and across the channel direction:

1. Lengthwise open – those extruders in which the material has a path open from inlet to outlet moving from the channels of one screw to the channels of the other
2. Crosswise open – those extruders in which there is in the area common to the two screws a path across the flights so that the material moves from a channel of one screw to two different channels of the other

It is, however, very difficult to decide to which category a twin-screw extruder belongs, as the flow path of the material depends not only on the direction of rotation of the screws (corotating or counter-rotating) but also on the shape of the flights and channels. The geometry of the screws has first to be thoroughly examined.

It seems better, therefore, to categorize these extruders according to the geometry of the screws, flights, and channels; the path of the material comes as a consequence of this geometry.

The most important basic subdivision depends on the position of the screws in relation to one another. There are extruders with:

1. Nonintermeshing screws in which the screws are next to each other, but as the name implies, they do not mesh or engage each other's threads, which are only tangent
2. Intermeshing screws in which the screws are also next to each other but the distance between their centers is less than twice their outside radius and, therefore, the flight of one screw

penetrates the channel of the other, or one screw engages with the other

The second subdivision concerns the shape and size of the flights and channels of the screws:

1. Nonconjugated screws are those in which the flights are such as to fit loosely into the channels of the other screw and leave ample passages all around.
2. Conjugated screws are those that have flights that have the same shape and dimension of, and tightly fit the channels of the other screw with a minimum of clearance.

The direction of rotation of the screws, which has little effect in nonintermeshing screws, assumes great importance when the screws intermesh, as we will see. Extruders with intermeshing screws must therefore be divided into two groups:

1. Corotating, in which the two screws rotate in the same direction, both clockwise or both counterclockwise
2. Counterrotating, in which the two screws rotate in opposite directions, one clockwise and the other counterclockwise.

Apart from "direction of rotation," which is clear, two words need clarification, as sometimes they are improperly used:

1. "Intermeshing" is by definition to engage together; two screws can engage one another fully, partially, or not at all, depending on how deep the flights of one screw penetrate the channels of the other. Similarly, two gears engaging their teeth are fully intermeshing because their teeth enmesh for their full height and could be partially intermeshing if they are not close enough for their teeth to totally conpenetrate.
2. "Conjugated" means mated; two parts are conjugated when the projections, or parts protruding from one, perfectly fit into the cavities of the other.

The use of "partially intermeshing" to indicate intermeshing nonconjugated screws, although common, is somewhat misleading and should be avoided.

With reference to the Erdmenger classification and according to the above definitions, we can divide twin-screw extruders as follows:

Screw Engagement	Flight and Channel Profiles	Counterrotating	Corotating
Fully intermeshing	Conjugated	Lengthwise closed Crosswise closed	Lengthwise open Crosswise closed
	Nonconjugated	Lengthwise open Crosswise closed	Lengthwise open Crosswise open
		Lengthwise open Crosswise open	Lengthwise open Crosswise open

According to what category a machine belongs in, it will work in a very different manner, as it is the difference in the geometry of the screws that makes a basic difference in their *modus operandi* (Fig. 2-2). Another type of twin-screw extruder that seemingly needs to be placed in a separate category is the one with conical screws. This type, however, does not operate any differently than any other twin-

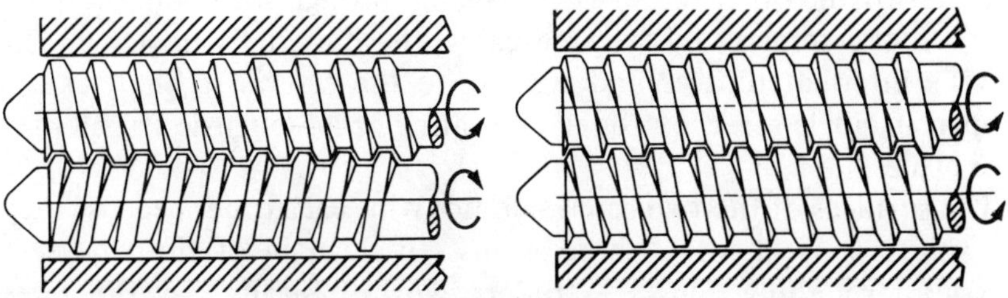

These twin screws rotate opposite to one another. In this case, the screws rotate in the same direction.

Fig. 2-2. Counterrotating and corotating screws arrangement.

screw machine. The interaxis and the diameter of the screws will have to be integrated along each segment of the screws, but all relationships and calculations for intermeshing twin-screw machines can otherwise be used for them. These extruders will therefore fall into one of the categories enumerated above.

There can be, of course, all kind of intermediate solutions such as "partially intermeshing" screws, where the screws intermesh but the center-to-center distance is larger than the outside radius plus the root radius of the screws, or "quasi-conjugated" screws, where some but not too large passages are left. These are compromise versions that have some of the advantages but mostly the disadvantages of both systems.

PRINCIPLES OF OPERATION

Machines with nonintermeshing screws operate very similarly to single-screw extruders. Along the periphery of each screw, the relationship between the screw, the barrel, and the material is the same, and the frictional coefficient of the material on the metal is the main factor controlling extrusion.

Where the screws are tangent, the material has a free passage from one screw to the other, but nothing really forces the material to move from one screw to the other, and in these machines neither pumping nor mixing is a positive action.

High backflow is always present, as the situation resembles the one in single-screw extruders. Under certain conditions of temperature, head pressure, and viscosity, the pumping and mixing are good, and the shallow flights promote good shear; however, under some other extruding conditions, the material within the channels of one screw may not mix but only "rub shoulders" with the material filling the channels of the other screw when passing through the common space between the screws. In extreme cases, the material will slip on the barrel's internal surfaces and stick onto the screws so that, as can happen in single-screw extruders, pulsating or no extrusion at all may take place.

The surfaces of the two screws do not wipe each other because they are apart; consequently, there is no self-cleaning of the channels.

Nonintermeshing screw extruders work basically on the same principle as single-screw extruders; they usually work well, but friction is still the "prime mover," and if there is no friction, there is no extrusion. These machines should therefore be put into a category separate from other machines with two screws and more appropriately called "double-screw" extruders.

Machines with intermeshing screws, or the real twin-screw extruders, do not work the same way. In them, there is an actual interaction of one screw on the material contained in the channels of the other screw when, in the intermeshing region of the flights, one screw actually penetrates the channels of the other screw. It is this interaction that makes this kind of extruder so different from any other.

Where the screws intermesh, the very presence of the flight of one screw within the channel of the other limits the rotational motion of the material around each screw so that forward motion is achieved (Fig. 2-3).

Machines with intermeshing screws have even better propulsive characteristics when they have conjugated profiles. In these machines, where the screws meet and intermesh, the channels are not only restricted but also sometimes completely closed by the flights of the other screw, thereby totally impeding the rotation of the material around each screw. The material cannot rotate together with the screw, not even if it tends to stick to it, for it is forcibly stopped by a metal wall (Fig. 2-4).

With the rotation of the screws, the material contained in the channels is instead forced to proceed axially forward along the barrel. This action is positive and not dependent on the operating conditions (type of material, temperature, pressure, etc.), but due only to the geometrical characteristics of this type of machine.

This is the basic difference between single- (or double-) screw extruders and real twin-screw extruders. In these latter, pumping action is positive, and so is the mixing; the more the profile of the

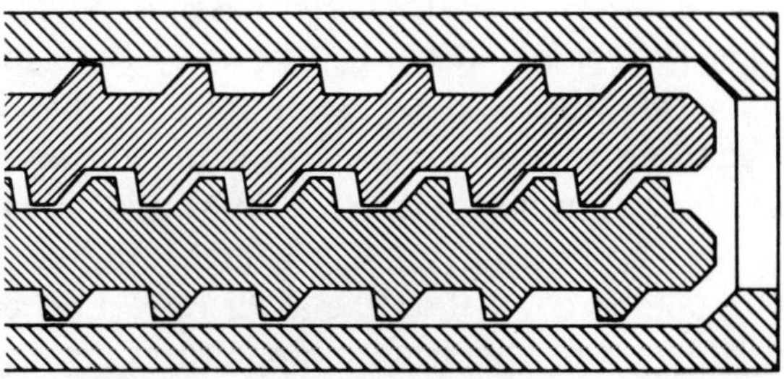

Fig. 2-3. Longitudinal section of corotating intermeshing nonconjugated screws showing some passages around each screw.

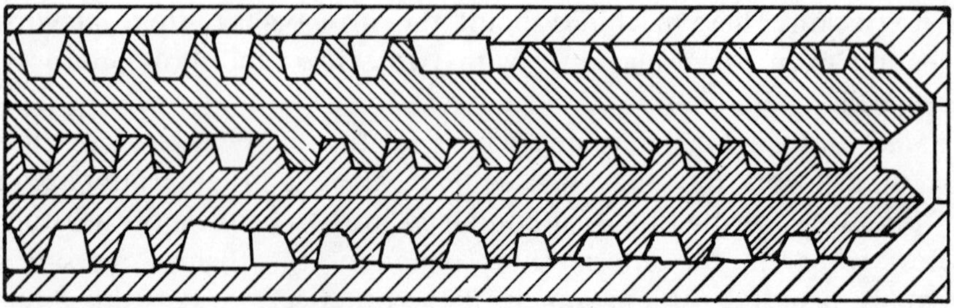

Fig. 2-4. Longitudinal section of corotating intermeshing conjugated screws (Colombo) showing closed passages around each screw.

flights matches the profile of the channels or the better the screws are conjugated, the more these interactions are preeminent and positive.

Whether rotating in the same or in opposite directions, extruders with intermeshing screws work on completely different principles than single-screw or nonintermeshing double-screw extruders; therefore, only to machines with intermeshing screws should be applied the nomenclature of twin-screw extruders.

In proper twin-screw extruders, the screws act, more or less, like a positive displacement pump and depend only minimally on the friction of the material against the barrel to move it forward.

Disregarding mechanical clearances, counterrotating screws may have no passages at all for the material to move around each or both screws, so that it must move axially toward the exit.

Corotating screws may have no passages at all for the material to move around each screw and only but small passages for it to slowly move around both screws. The smaller these passages, the more positive the propulsive action of these twin-screw extruders.

The L/D ratio is of nonsignificant importance for the propulsion of the material because conveying action is not much affected by difference in head pressure, so there is no need for a long metering zone. Second, in this type of machine, usually the material is fed by a separate mechanism that measures it into the extruder; therefore, there is no fixed relationship between output and screw speed, and this independence, permitting relatively high output at low screw speed, allows for a good control on the shear rate. Last but not least, with these extruders, the operating temperatures do not have to be restricted to those that keep the frictional coefficient of the material within adhesive range, nor will a small variation in temperature affect the propulsive action with consequent pulsations.

In intermeshing screws with conjugated profiles, however, the direction of rotation of the screws, whether corotating or counter-rotating, has great influence on the way these extruders work and operate.

INTERMESHING AND CONJUGATION

From what has been said above, it is obvious that the most important point in twin-screw extruders is where the screws intermesh. To clearly understand the working principles of twin-screw extruders, the geometry of the flights in the intermeshing region has to be thoroughly examined first. Consider, for a moment, having screws with rectangular flights similar to the extrusion screw of a single-screw extruder but with very deep and relatively large channels and very high flights of negligible thickness. Then we place two of these screws next to each other so that they fully intermesh.

If the diameter of each screw, which we will call the primitive diameter, is D', and they are placed at a distance I equal or slightly larger than $D'/2$ from each other (interaxis), the depth of intermeshing will be $h' = D' - I$, and this will be the height of the flights and the depth of the channels (Fig. 2-5).

When the two screws penetrate each other's threads, their flights have to pass freely within the channels of the other screw. If we select the flights of each screw, as we just described, when the screws rotate,

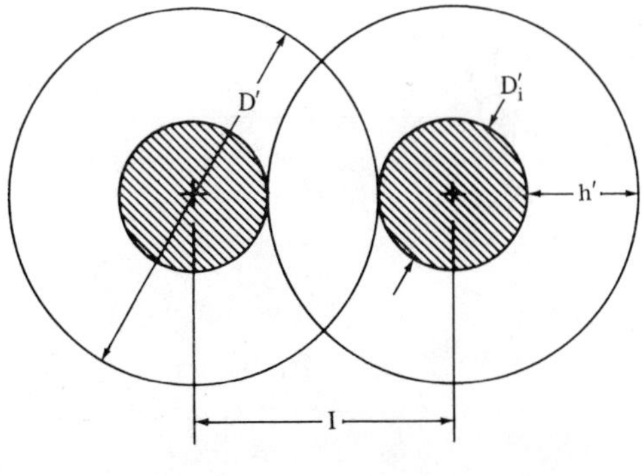

Fig. 2-5. Cross section of intermeshing twin screws.

these flights still have to be able to move without interfering with each other.

Figure 2-6 depicts, in prospective view, a section of the two screws. In it is represented a section of the two screws' outside diameter, on the same place with centers O and O'; the area included in the arc of circumference between the points A and C is the zone of intermeshing. The other lines' departing point A are the traces of the spiraling tip of the flight in the case of the left-hand thread of the far screw (A–F) and right (A–D) and left (A–E) thread of the near screw.

On the far screw, the spiral A–F would push the material toward the left of the drawing when the screws rotate counterclockwise. On the near screw, the spiral A–D will also push the material toward the left of the drawing when turning clockwise. The spirals A–F and A–D, therefore, mark the paths of the flights of each of two counterrotating screws. On the near screw, the spiral A–E will push the material to the left if the screw rotates counterclockwise. The spirals A–F and A–E mark, therefore, the paths of the flights in two corotating screws.

This shows that in intermeshing counterrotating screws, the flights of each screw, which are next to each other at the position A, move

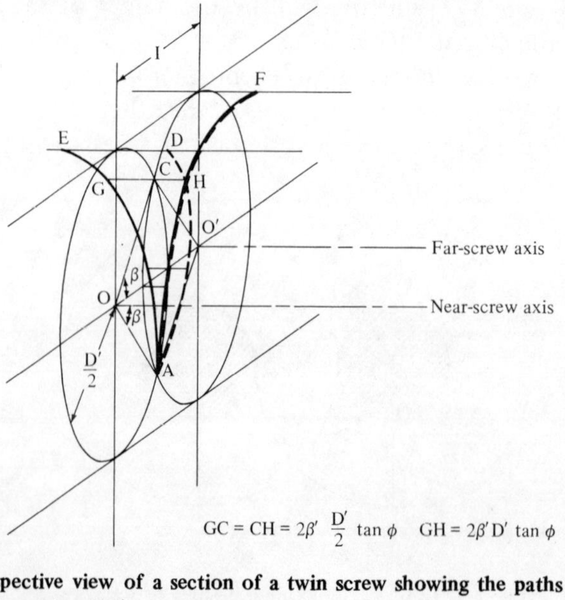

$$GC = CH = 2\beta' \frac{D'}{2} \tan\phi \qquad GH = 2\beta' D' \tan\phi$$

Fig. 2-6. Prospective view of a section of a twin screw showing the paths of the flight in corotating and counterrotating screws.

along their respective paths remaining on the same plane until, at position *H*, before they leave the intermeshing region, they are still next to each other. The plane on which these points move is inclined in relation to the screw by the pitch angle ϕ, but each point of each screw moves on a plane inclined by the same amount, therefore never interfering. Any type of flight (rectangular or other shape) can thus be used.

In the intermeshing region, the flight of one screw enters the channel of the other screw in the middle of it and, when rotating, stays at equal distance from the flanks of the channel until it leaves it.

The flight will always remain centered in the channel, and it can be made as thin as mechanically feasible or as thick as the channel is wide.

When the flights are as thick as possible, where they intermesh, they will not leave any passage between the screws, as they are perfectly conjugated on the plane of the screws' axis. The material remains enclosed in C-shaped chambers around each screw, thereby enhancing the pumping but drastically reducing the mixing action. As the screws revolve, the material is shifted axially toward the extrusion end of the barrel, like a nut on a bolt, without having a chance to mix with the material contained in the other channels. Within this closed C-shaped chamber, the material is dragged by the rotating screws toward the point where the screws meet.

There it is pushed from both sides, and at this point, where the screws meet, its pressure greatly increases. As there is no other way for it to go, it squeezes into the intermeshing region through mechanical clearances, moving from the top to the bottom of the screws (or from the bottom to the top, according to the direction of rotation, both screws clockwise or both counterclockwise), thereby undergoing high shear (Fig. 2-7).

Furthermore, the same pressure exercises a force on the screws that tends to push the screws apart and make them rub against the barrel.

If the flights are made thinner than the channels, although the screws intermesh, there will be no conjugation. The flights of one screw only minimally disturb the material in the channels of the other screw and only partially impede its rotational motion around each screw, thereby somewhat defeating the twin-screw principle. Rectangular flights (and channels) therefore present difficulties and a different shape of flights will give better results.

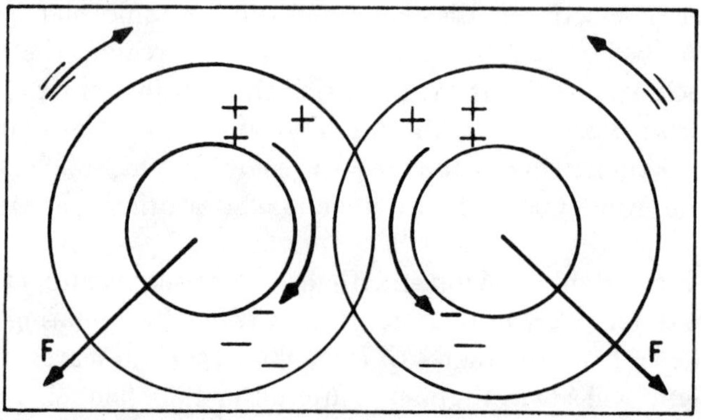

Fig. 2-7. Pressure produced in counterrotating screws by drag flow.

As the screws in the intermeshing region have the flights inclined to an angle ϕ but move parallel to each other, any shape of flight can be chosen without fear of interference. A type of flight that has some advantages has a trapezoidal shape, or it is wider at the base than at the tip. To make the screws conjugate, the channels in this case must be made conversely wider at the screw's outside diameter and smaller at the bottom (Fig. 2-8).

This type of flight, though perfectly conjugated at the plane of the screw's axis, does leave some passages for the material to move from one screw to the other. A rectangular flight, perfectly conjugated, must have a flight width E equal to $p/2$; a trapezoidal flight with a flight tip width e smaller than E, when entering into the wider channel, will leave two passages leading to two different channels, each of a width equal to $p/2 - e$. This shape, on the one hand, satisfies the basic principle of twin-screw machines as they are conjugated,

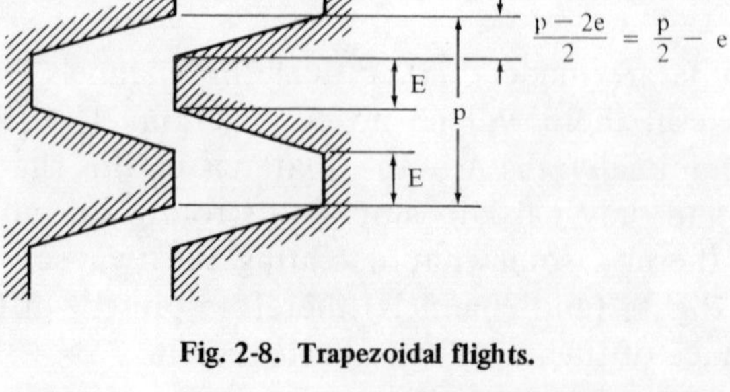

Fig. 2-8. Trapezoidal flights.

therefore the material cannot rotate with the screws. On the other hand, it positively pushes the flows of material around the screws to mix with one another through the abovementioned passages.

It remains a fact, however, that the converging drag flow presses the material on one side of the screws, mixes it through those passages somewhat, but still squeezes it through the clearances toward the other side of the screw's axis plane where a zone of low pressure is created by the same drag flow. To avoid the consequent high shear, compromises between small clearances, perfect conjugation, self-cleaning, and larger clearances with less shear but loss of conjugation have to be made.

In intermeshing corotating screws, the situation is greatly different. The flights of one screw are next to the corresponding flights of the other at the position A, but once the screws have rotated of the angle β', one flight is gone forward and the other rearward so that they become axially separated. After a rotation of $2\beta'$, at the end of the intermeshing area, they will be separated by the distance $\overline{GC} + \overline{CH} = \overline{GH}$, which can be measured as $2\beta' D' \tan\phi$.

In this kind of machine, screws with rectangular flights similar to what are used in single-screw extruders cannot have conjugated profiles, as the channels must be larger than the flights.

A section made on the screws along the lines OAO' showing the flights just engaging each other, with the flights of one screw close to one side of the flights of the other screw, is shown in Figure 2-9, which also shows sections made along OBO' and OCO'. Comparing the three sections, we can see that the edges X and X' of the flights move in relation to one another along the dotted line marked in section OBO'; this line makes an angle θ with the plane perpendicular to the screw's axis.

If we now want to thicken the flights and make them so that they have the same shape and completely fill the corresponding channels of the other screw, the flank of the flights has to be constructed inclined by the slope shown by that line, or with an angle θ, so that the edge X slides near but never touches or interferes with the other screw. The angle θ can therefore be called the sliding angle.

Following that guideline, we will have a triangularly shaped channel and a triangularly shaped flight with a flank angle equal to θ on each side.

With reference to Figure 2-9, we can see that $\tan\theta = p/2h'$. This form of flight, however, presents problems in extrusion such as the

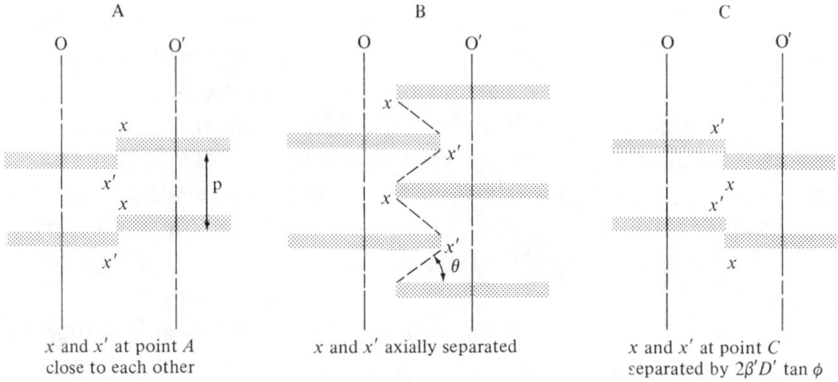

Fig. 2-9. Position of the screws' flights in relation to one another in corotating screws. (A) X and X′ at point A close to each other; (B) X and X′ axially separated; (C) X and X′ at point C separated by 2β′D′ tan θ.

very small resistance that these pointed flight tips sliding against the barrel offer to head pressure and leakage flowing of the material.

Another form of flight that permits perfectly conjugated screws without interference is a sinusoidal shape of flights and channels, but this sinusoid must have the angle of the tangent at the flex point equal to the sliding angle. This form also has a minimal resistance to leakage flow between the barrel and the flight's tip where the sinusoid has a tangent parallel to the screw's axis.

In triangular flights, the problem of leakage can be avoided, giving the flights a flat land against the barrel by truncating the tips of the flights to a width E; the resulting cross-section of the flights will be trapezoidal.

It is evident that by cutting off the tips of the flights, in order to maintain the angle of the flanks equal to the sliding angle, we will have to reduce the screw's diameter from D' to D.

On the one hand, to provide for a free movement of the flights within the channels, as we have seen, we must have the flanks inclined by $\tan\theta = p/2h'$; on the other hand, cutting the tip of the flights to a width E, the pitch, which remains the same, can be expressed by $p = 2h \tan\theta + 2E$. We have then:

$$p = 2h \frac{p}{2h'} + 2E$$

The channel depth, which was $h' = D' - I$ will be $h = D - I$, so we can solve for D, and we have:

$$D = \frac{(D'-I)\ (p-2E)}{p} + I$$

which is the new diameter of the screws when the primitive diameter D' is reduced by truncating the tip of the flights to a width E.

If D' was chosen equal to $2I$, as is usually the case, D would become

$$D = \frac{I}{p}(p - 2E) + I \quad \text{or developing} \quad D = 2I\,(1 - \frac{E}{p}).$$

The width of the tips E can be chosen between $E = 0$ (triangular flights) and $E = p/2$. In this latter case, however, D will be equal to I and therefore h will be equal to zero; the screws will no longer intermesh and will not even be screws anymore but straight cylinders.

Within those extreme limits, any screw with the newly calculated diameter D for the flight-tip width chosen will have the flanks inclined at an angle θ and will be perfectly conjugated, still not interfering with each other. Usually, E is chosen between 1/3 and 1/4 p.

These limitations in the shape of the flights, dictated by geometry, both in counterrotating and in corotating screws, are the reason for the great difference in the way these two groups of extruders operate.

We can briefly summarize these differences to show how the direction of rotation influences the construction and thereby the manner of operation of twin-screw extruders.

In intermeshing twin-screw extruders where the screws are conjugated and rotate in opposite directions, the material, while it is transported by each screw, clockwise or counterclockwise, to the point where the screws intermesh, whatever the shape of the flights, finds closed or almost closed channels. This creates a positive plug flow toward the die; however, at the point where the two opposite flows converge, there tends to be an accumulation of material and the consequent creation of high-pressure zones.

By the same token, low-pressure zones are created on the opposite side where the channels part. The pressure varies from point to point along the periphery of the screws with a maximum where the material converges. Resultant forces tend to separate the screws and press them against the barrel, with consequent wear. The material, which is pushed out of the channel of one screw by the flights of the other, is squeezed between the screws from the high-pressure to the low-pressure zones, and the laminating action and the high velocity of the material at this point develop localized high shear. To minimize this effect, conjugation is given up, and gaps or large clearances are left between the screws or purposely cut in the flights. On the other hand, if these passages are large, the propulsive ability is not only greatly reduced by the increased backflow, but also it becomes less positive, making the machine more sensitive to variation of head pressure.

Mixing is determined mainly by the size of these gaps or purposely cut passages; the closer they are, the more the material is held in the C-shaped channels around each screw without shuffling or mixing until it reaches the end of the barrel.

In counterrotating twin-screw extruders, mixing, pumping, self-cleaning, as well as the amount of heat transferred to the material depends on the shape and size of these passages. Whereas a given geometry favors certain aspects, the same geometry simultaneously creates disadvantages in others.

If the screws are rotating in the same direction, either both clockwise or both counterclockwise, the material is transported by the screws toward the die. In this case, as we will see better later, the material pushed out of the channel of one screw by the flight of the other moves into the channel of the second screw. At the intermeshing point, the material cannot proceed around the same screw, as the conjugated profiles completely close the passages. The flank of the screw's flight penetrating the channel acts as a wedge and forces the material to leave that channel and move into adjacent channels of the other screw. The transfer of material from one screw to the other creates a movement around both screws. The better the profiles are conjugated, the slower this movement and the greater the propulsive action.

Corotating twin-screw extruders do not tend to accumulate material at any point around the screws, and pressure is the same all around. There is no squeezing of the material with consequent possible overheating; there are no forces to push the screws apart. The material itself keeps the screws centered in the barrel; for this reason, close tolerances can be kept between the screws and the barrel as within the screws. This means that the screws will cleanse each other completely and rapidly, without wear of screws or barrel.

This self-cleaning action is of great importance, as it avoids hangup of material in any place along the screws. With some types of highly sensitive material, particles that remain stuck on the screws' surfaces, especially at the bottom of the channels, will slowly decompose if they are not wiped away.

Furthermore, this cleaning action, together with the positive pumping, moves the material within the screws, making it all advance at practically the same speed, thus equalizing residence time for all particles. Last but not least, the self-cleaning allows for a neat and rapid change of the material's color, without streaks or leaving behind particles that may break loose at any time, later marking the new material with spots of the previous color.

GEOMETRY OF SCREWS

The type of extruders we are going to examine in more detail are the true twin-screw extruders, either counterrotating or corotating.

Our analysis will be more conceptual than mathematical in order to explain the functioning, the differences, and the choices available. We will use very simple elementary formulas for a more qualitative than quantitative analysis of the various phenomenon examined.

According to old habit born of single-screw extruders, the system of classifying extruders by their screw's diameter has been usually

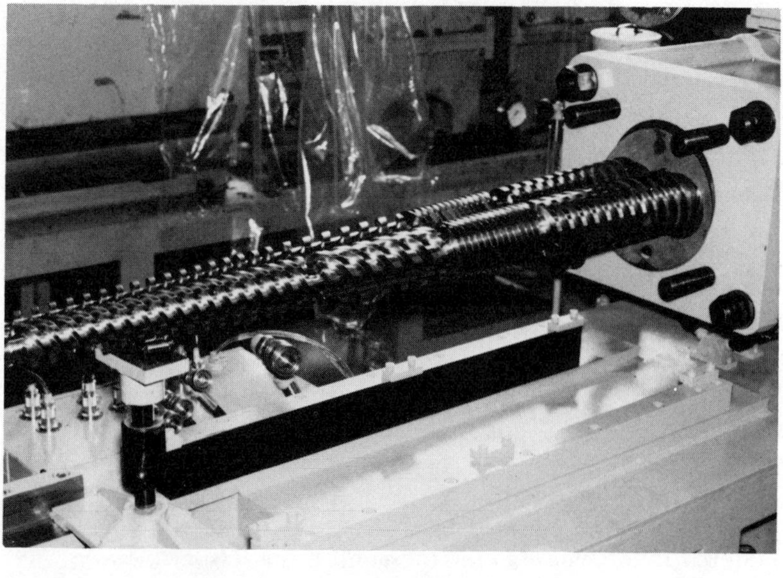

Fig. 2-10. Counterrotating conical screws for PVC compounding (Toshiba).

Fig. 2-11. Counterrotating screws (Reifenhauser Nabco).

kept also for twin-screw extruders. In these extruders, however, a screw's diameter is not the characterizing factor; from the screw's diameter alone we could not possibly know how much intermeshing there is or how deep the channels are. These factors are also important to know to characterize a twin-screw extruder.

The first and most important parameter is the center-to-center distance of the screws. The basic relationship in any twin-screw extruder is:

$$I = \frac{D}{2} + \frac{D_i}{2} \quad \text{or} \quad D + D_i = 2I$$

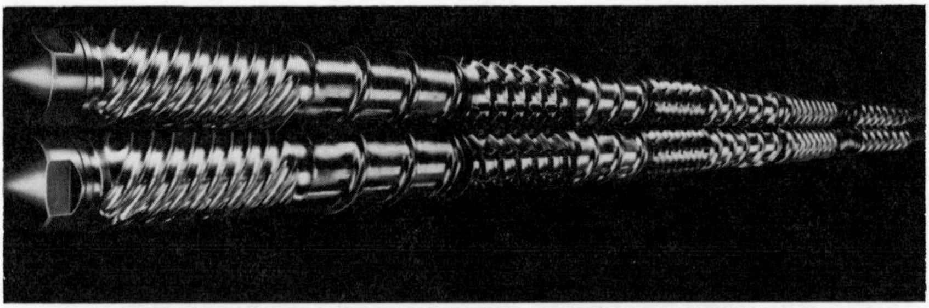

A

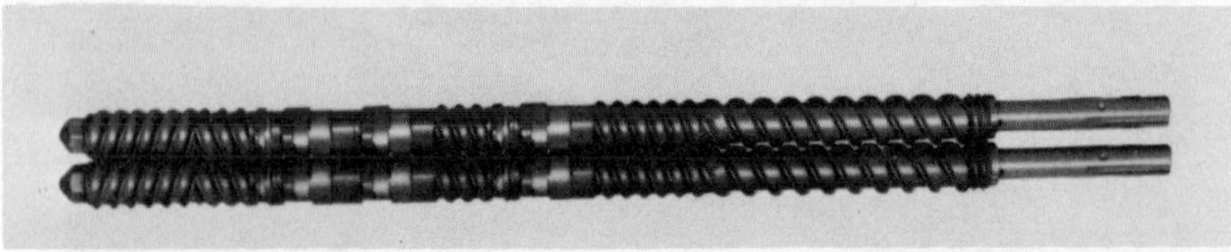

B

Fig. 2-12. A. Corotating screws (Berstorff). B. (Werner and Pfleiderer).

where I = center-to-center distance or interaxis, D is the screw's external diameter, and D_i is the root diameter.

In order to have intermeshing, the values of D and D_i are restricted so as to satisfy:

$$I < D \ll 2I; 0 \ll D_i < I$$

We can immediately derive from the above the depth of the channels:

$$h = \frac{D - D_i}{2} = \frac{D}{2} - (I - \frac{D}{2}) = D - I$$

Interaxis alone, therefore, gives a very good idea of all the basic dimensions we have to know in order to characterize the extruder.

So, for example, a 3″ interaxis extruder must have screws with a diameter of more than 3″ and much less than 6″, or somewhere around a 4″ diameter; it also must have the roots of the screws smaller than 3″, or around a 2″ diameter; the depth of the flights must be around 1″.

Of course, the interaxis alone is not enough to calculate the other dimensions, but a good "educated guess" can be made so as to have an idea of the characteristics of the extruder in question.

Interaxis is the dimension to start from in the project and design of twin-screw extruders, as this dimension controls:

1. the shafts and gear size, therefore the torque available at the screws
2. the thrust-bearing size, therefore the maximum head pressure
3. the screw's external and root diameters, therefore the amount of intermeshing and the depth of the channels.

Another important parameter that has to be established in order to define the geometry of the screws is the angle of the screw's elix ϕ. This is the angle that the flight makes with a plane perpendicular to the screw's axis while it winds around the roots of the screw.

This angle ϕ determines the pitch of the screw or the distance from flight to flight measured in a direction parallel to the screw's axis.

Most single-screw extruders have a pitch lead equal to the diameter of the screw, which results in the so-called square pitch. This relatively long pitch and the small height of the flights give a certain volume to the channel; in twin-screw extruders, where the channel's depth is much higher, in order to maintain the channel's volume within reasonable limits, the pitch is much shorter; usually, the pitch reaches three or even four leads per diameter (Table 2-1). The angle ϕ is therefore much smaller than the 17.76° required for a square pitch and usually varies around 10°.

The pitch is usually maintained constant along the whole screw, or in the more common case of screws made in separate segments, it is constant for every segment. For segmented screws, the pitch varies from 15° in the rear to 6° or 8° in the last section.

This very short pitch has the added advantage that the rotational component of the pumping action is very small; therefore, the torque required to generate the necessary extrusion pressure is much smaller with the short than with the longer pitches. Furthermore, short-pitch flights provide a better barrier and hold the material under pressure much better because they are closer to being perpendicular to the screw's axis.

We can easily see that the pitch is:

$$p = \pi D \tan \phi$$

and also that the length of the elix formed by the flight is:

$$S = \frac{\pi D}{\cos \phi}.$$

Whatever the shape and size of the flight, in order for the screws to be conjugated, the channel must have, apart from mechanical

Table 2-1. Elix Angle and Resulting Number of Pitches Per Diameter.

ϕ°	PITCH	PITCHES PER DIAMETER
17.76	1.00D	1.00
15	0.84D	1.18
10	0.55D	1.80
8	0.44D	2.26
6	0.33D	3.00
5	0.27D	3.60

tolerances, the same shape and size. Whatever that shape is, the area of the channel will be one-half the rectangular area, having as sides the pitch p and the channel depth h; the other half will be exactly the same, being the area of the flight.

The channel area is:

$$A = \frac{ph}{2} = \frac{1}{2}\,\pi D h \tan\phi$$

which, multiplied by $\cos\phi$, gives us the area of the channel measured in a direction perpendicular to the channel itself. We have, then:

$$A = \frac{1}{2}\,\pi D h \sin\phi.$$

In conjugated screws, the material cannot stay within the channel of one screw and rotate around it because, at the intermeshing, in the plane of the screw's axis, these channels are filled by the flights, and the passage is blocked. In other terms, the material cannot "stick" to each screw, but it must move in relation to it. Friction and frictional coefficients variable with temperature, which are the main reason for the motion of the material in a single-screw extruder, do not play any major role here. Twin screws have the inherent capacity of forcing the material to advance and move toward the exit.

To calculate the volume transported by the screws, we cannot think of a cross-section A moving with a speed V along the channel, as the flow within the channels is not continuous; as a matter of fact, it is restricted in the intermeshing area and totally closed at the plane of the screw's axis.

Where the two screws intermesh, the flight of one screw, by definition, enters the channel of the other screw. The angle of intermeshing, defined as shown in Figure 2-13, is given by trigonometry as $\cos a = b/c$, so that we can write:

$$\cos\beta = \frac{I/2}{D/2} = \frac{I}{D}$$

The channels of the two screws, if we do not consider the parts where they intermesh with each other, are shaped like two big C's facing each other. In counterrotating twin-screw extruders, these chambers are practically closed, and we can calculate their volume.

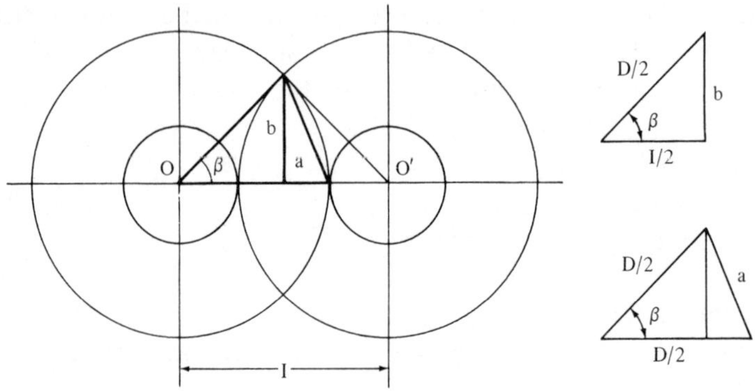

Fig. 2-13. Geometry of the intermeshing area.

To determine the length of the C-shaped chambers, we will deduct from the circumference of each screw the intermeshing part, and the remaining length is:

$$C = \pi D - 2a.$$

We can calculate a in an approximate way by solving the second triangle in Figure 2-13, and we can write:

$$a = \sqrt{\left(\frac{D}{2}\right)^2 + \left(\frac{D}{2}\right)^2 - 2\left(\frac{D}{2}\right)^2 \cos\beta} = \sqrt{\left(\frac{D}{2}\right)^2 (2 - 2\cos\beta)}$$

substituting $\cos\beta$ and because $h = D - I$, we will have:

$$a = \sqrt{\frac{1}{2} Dh} \text{ and } 2a = \sqrt{2Dh}$$

So we can now write the length of these chambers as:

$$C = \pi D - \sqrt{2\,Dh}$$

and their volume:

$$A \times C = \left(\frac{1}{2}\,\pi Dh \sin\phi\right)(\pi D - \sqrt{2Dh})$$

In corotating screw extruders, however, these are not closed chambers, but from the point where the screws begin intermeshing to the plane of the screw's axis, the combined flights form passages that lead from one screw to the other.

In fact, at the beginning of the intermeshing, the flights of the two screws are next to each other, as shown in Figure 2-9, section OAO'; if we make a similar section, when the flights are trapezoidal, we can see that it will look as in Figure 2-14, leaving a passage between the two channels. The material can move from one screw to the other, which rotates in the same direction.

Whereas no passage is left in the plane OBO', along the line $\overline{AB}$ there is a triangular opening that, in the direction of the screw's axis, has a width $p - 2E$ at the point A, and a width zero at the point B. Its height $\overline{AB}$ can be calculated as $\frac{1}{2}\sqrt{D^2 - I^2}$. This passage reaches the channel of the other screw, and there it has the same triangular opening but inclined in relation to the first so as to define the solid form of a tetrahedron with slightly curved sides.

The dimension of this tetrahedron is shown in Figure 2-15, and as can be seen, it provides a way for the material to be transferred from the channels of one screw to the channels of the other.

Although these passages do not have the same area as the channels themselves, they connect the two otherwise closed C-shaped channels and form a single continuous channel in the shape of a twisted figure 8 (Fig. 2-16).

To determine the length of this figure-eight-shaped channel, we will deduct from the circumference of each screw the intermeshing part. The remaining length is:

$$C_e = 2\,(\pi D - \sqrt{2Dh})$$

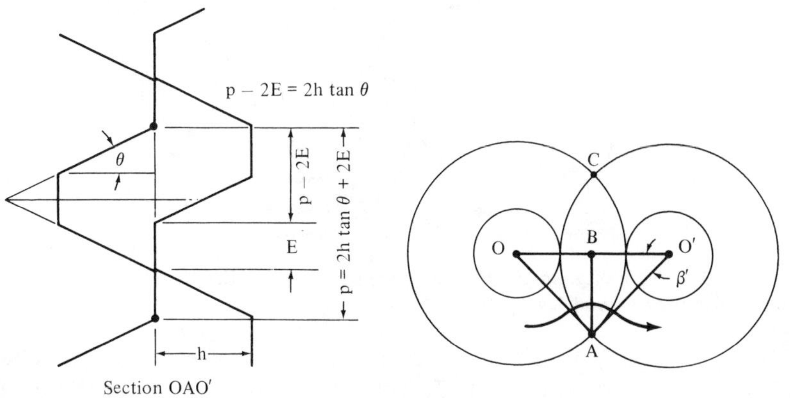

Fig. 2-14. Cross sections of corotating intermeshing conjugated screws with trapezoidal flights.

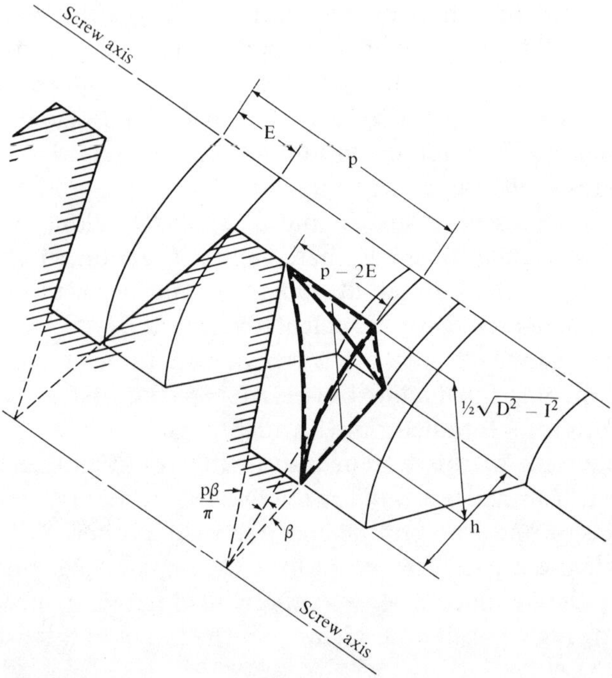

Fig. 2-15. Passage in the shape of a tetrahedron left between corotating intermeshing screws by trapezoidal flights of tip width E.

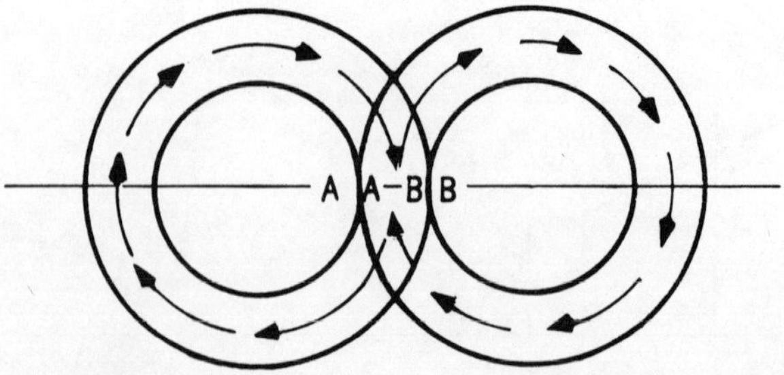

Fig. 2-16. Schematic diagram of corotating twin screws shows figure-eight path of material. Material cannot pass through section A-B.

This length is equal to the circumference of a screw having a diameter of:

$$D_e = \frac{2}{\pi}(\pi D - \sqrt{2Dh}) = 2D - 0.9003\sqrt{Dh}$$

This hypothetical screw with a diameter D_e and a circumference C_e can be called the "equivalent screw."

This "equivalent screw" has each turn of the elix as long as the length of the figure-eight-shaped channels around the two screws. The length of this channel can be written:

$$S_e = \frac{\pi D_e}{\cos\phi} = \frac{\pi}{\cos\phi}(2D - 0.9\sqrt{Dh})$$

In the example previously given, in a corotating twin-screw extruder with I=3″ or with two 4″ diameter screws, the material moves around the figure-eight-shaped channels as around a 6.2″ equivalent-diameter screw.

The next important parameter after I, D, D_i, and ϕ is the width of the flight tip.

FLIGHT-TIP WIDTH

In counterrotating screws, the shape of the flight can be chosen almost at will, and still perfect conjugation can be kept, provided, of course, that the flights and channels have the same shape. A reduction of the flight-tip width can also be made without any geometrical interference between flights, with the result of having trapezoidal flights. This does not basically change the flow of the material; however, a reduction of the screw's tip width without corresponding reduction of the width of the bottom of the channel, which, of course, makes the screw lose conjugation, partially changes the flow of material.

With reduced flight-tip width and large channel bottom, large passages are created that allow the material to flow freely from the high-pressure to the low-pressure side of the screws. This movement greatly enhances the mixing of the material between the two screws, but with the creation of these openings, the pumping ability of the screws is greatly reduced.

Reduction of flight-tip width with loss of conjugation is therefore used in counterrotating screw extruders only for the mixing section of the screws, whereas good conjugation is kept where high pumping action is desired.

When the screws rotate in the same direction, the choice of the flight-tip width does change the flow of material, but without loss of conjugation.

We have seen before that in corotating extruders, E can be chosen at will, but between some stated limits, and once it has been chosen, we still have the possibility of making it actually equal to E or, if we want, to make it smaller than E, thereby causing the screws to work in a different way, without causing interference between the screws.

When we make the actual tip of the flight equal to E, the angle of the flight's flanks, which we call α, will be equal to the sliding angle θ; if we choose to make the flight-tip width e smaller than E, the angle of the flanks α will increase and will be $\alpha > \theta$.

When the tip of the flight has a width E, at the beginning of the intermeshing, it is close to the flight tip of the other screw, and when rotating, it slides near the flank until, on the plane of the screw's axis, it perfectly fits in the bottom of the channel of the other screw.

From Figure 2-17, we note that if we reduce the width of the flight tip, we still allow free motion of the screws without interference; also, at the beginning of the intermeshing, the flight tips of one screw are not close to the flight tips of the other screw; another passage is left, conducing to a different channel of the other screw.

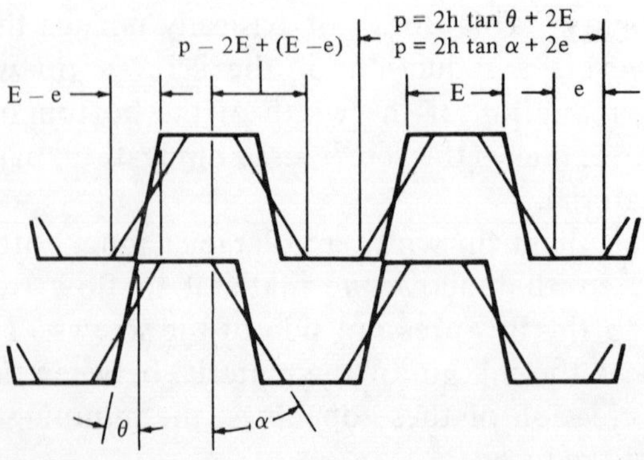

Fig. 2-17. Longitudinal section of corotating conjugated screws with flights tips E or $e < E$ made at the beginning of intermeshing.

The actual flight-tip width e can be chosen anywhere between zero and E; consequently, the actual angle α of the flight flanks, which should be more than θ for conjugation without interference, for $0 < e < E$, will be:

$$\frac{p}{2h} \geqslant \tan\alpha \geqslant \frac{p-2E}{2h}$$

and because $p = 2e + 2h \tan\alpha$, α must be such that:

$$\tan\alpha = \frac{p-2e}{2h}\ .$$

When the flight tip is equal to E, as we have seen, there is only one passage leading from a channel of one screw into a channel of the other; if, on the other hand, e is chosen smaller than E, the intermeshing flights remain conjugated but leave not one but two passages leading from one channel of one screw to two different channels of the other.

We have seen before the width of the single passage when the flight tip is equal to E; we can now see what is the widths of the two passages when the flight-tip width is less than E.

From Figure 2-17, we can see that when making the flight tip smaller than E, the passage W, already existing, will increase by $(E - e)$ becoming $(p - E - e)$, while the new passage W′ will have a width equal to the increase of W, or $(E - e)$.

If we make a longitudinal section of the screws in a vertical plane where the screws intermesh, we can very well see these passages (Fig. 2-18).

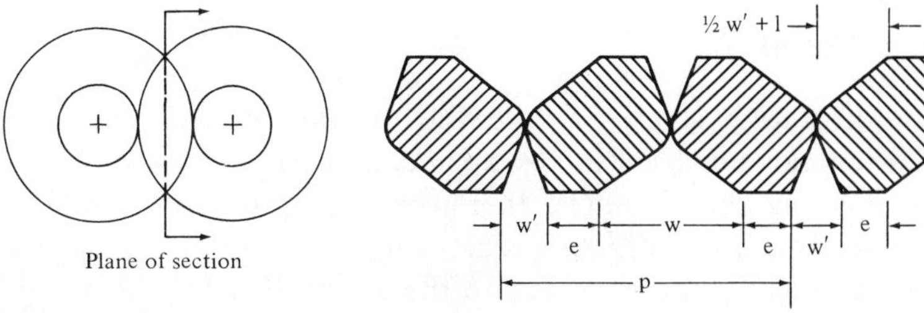

Fig. 2-18. Longitudinal section at the intermeshing area of corotating intermeshing screws with tip width E.

We can calculate the width of those all-important passages by calculating the axial shift of the flight with a pitch angle ϕ during its movement within the intermeshing angle β.

Going back to Figures 2-13 and 2-14, we have the length $b = \frac{1}{2} \times \sqrt{D^2-I^2}$ and for an angle 2β, the shift will be $l = 2b \tan \phi$.

Now considering, in Figure 2-18, that $\frac{1}{2} W' + l = \frac{1}{2} W$ and that $W + W' + 2e = p = \pi D \tan \phi$, by simultaneous solution, we have:

$$W = \frac{1}{2} \tan \phi \left(\pi D + \sqrt{D^2 - I^2}\right) - e$$

$$W' = \frac{1}{2} \tan \phi \left(\pi D - \sqrt{D^2 - I^2}\right) - e$$

and the sum of the two:

$$W + W' = \pi D \tan \phi - 2e = 2h \tan \alpha$$

The angle of the flanks α therefore characterizes the area of these passages. In actual measurement and according to the e chosen, we note that this area varies between 80% and 25% of the channel's area.

Having the means to choose the sizes of these passages without losing conjugation permits us to make corotating intermeshing screws particularly designed for mixing or pumping.

In the first case, most of the material contained in a chamber rotates around the two screws; in the other, even allowing for some rotation, most of the material is retained in the C-shaped, but open chambers and is forcibly propelled forward.

Usually, the area of these passages is 55–70% of the area of the channels in the feed and mixing zones and around 25–45% in the pumping sections.

SHEAR STRESSES

In counterrotating screws, where the profiles are conjugated, the material remains enclosed in C-shaped chambers, and there is no motion of it along the channels; therefore, a plug flow occurs. In corotating screws, however, even when they have conjugated profiles, there are passages from one screw to the other. If the flight-tip width is E, that is, as large as possible, as we have seen, there is one passage; if it is smaller, there are two. In any case, the material flows from the channels of one screw to the channels of the other, and the material moves around the "equivalent screw."

That the material does or does not flow in lengthwise open channels creates a basic difference in velocity and stress distribution within the channels. These differences are clearly shown in an excellent study by Werner & Pfleiderer. Without entering into the differential calculus used by this study, it shows: (1) In counterrotating twin-screw extruders, the shear stresses near the barrel and at the screw's root are greater than in corotating screws, so that in counterrotating screws these layers of material are greatly mistreated, whereas a large percentage of material inside the channels is not exposed to enough shear for good plasticization and dispersion. (2) In counterrotating screws, the velocity distribution along the channels, which varies greatly according to depth, can only be influenced by a change in screw pitch, which, however, is preselected. (3) In corotating screws, also, throughput and screw speed, which can be changed during extrusion, considerably alter velocity distribution; thereby, more uniform shear stresses can be achieved, with consequent better material homogeneity.

Furthermore, in counterrotating screws, the material forced into the calendering gap undergoes, as it has been mentioned before, a high shear but only a small fraction of the chamber's volume passes through it. This adds to the nonhomogeneity of the extrudate. Mixing devices as interrupted flights, etc., will ameliorate the results, although only partially.

In corotating screw systems, the very presence of the area restriction and shape change in the passage between screws further enhances homogeneity without localized high shear at any point.

OUTPUT

If we suppose that the material within the channels "slips" on the barrel surface, a fact that would stop extrusion in any single-screw extruder, obviously the material that fills each channel remains motionless within it (in relation to it) and rotates with the screws. In twin-screw extruders with conjugated profiles, once the material rotating with the screws reaches the intermeshing area, it can no longer rotate with them because, since the profiles are conjugated, the flights of one screw completely fill the channels of the other screw, thereby closing the way around the screws. The material therefore must stop its revolving motion, and as the screws keep rotating, it slides within the channels, assuming an almost linear motion toward the end of the barrel. This is positive pumping.

When the screws rotate in opposite directions, as we have seen, the material is more or less sealed into chambers formed by the intermeshing conjugated threads. The material must remain in these separate chambers and, as the screws rotate, must advance forward.

The maximum possible output of these types of extruders would be, therefore, per each screw, equal to the volume of one of these chambers, already calculated before, by the rpm of the screws. For two screws, we can write:

$$Q_c = 2\, i\, V\, n$$

where i is the number of thread starts and n is the number of revolutions per second (rps) of the screws.

For a single lead screw, that results in:

$$Q_c = \pi D h \sin \phi \, (\pi D - \sqrt{2Dh})\, n$$

These machines are very insensitive to the pressure generated by the die; therefore, in the equation of the output, there is not a term similar to the "back pressure" as in single-screw extruders. The output given by the formula, however, is not the actual output, as there is always a "leakage flow" around the screws. Furthermore, when the screws are not conjugated or when the flights have a form that does not completely fill the channels but leaves some passages, there is also a leak within the intermeshing region.

In counterrotating screws, the drag effect makes the material flow in opposite directions, so that, where the screws meet at the intermeshing, these flows also meet. The uneven pressure created within the channels by these counterrotating flows squeezes part of the material out of the C-shaped chambers. The larger the gap between channels and flights (for lower shear), the greater is this leakage, and therefore the greater is the reduction in the pumping efficiency of the machine.

For counterrotating screws, Menges and Klenk, as well as Doboczky, state an efficiency of .35 to .5, so that the maximum output becomes:

$$Q_{max} = \eta Q_c = Q_c - q$$

where $q = .65$ to $.5\ Q_c$.

Although an increase in rpm increases the theoretical maximum-output capability, it also increases the pressure differential around the screws, therefore increasing the speed of the material through the gaps, affecting its viscosity and decreasing the pumping efficiency. While Q_c increases linearly with n, the maximum output is only slightly affected by the screw's speed. So, for example, increasing the screw's speed from 10 to 30 rpm increases Q_c from 20 to 60 kg/hr; q increases also from 10 to 30 kg/hr, so that the real increase in output would be only from:

$$Q_c - q = 20 - 10 = 10 \text{ kg/hr}$$

to:

$$Q_c - q = 60 - 30 = 30 \text{ kg/hr}$$

showing a marked reduction in pumping efficiency.

In corotating twin-screw extruders, the material is not kept in close chambers, but as we have seen, there are passages left by the geometry of the screws.

When the screws are conjugated and the screw's flight tip is equal or close to E, those passages are minimized, and we have a propulsive screw that makes the material forcibly advance toward the barrel's end. The material flowing through these passages, however, does not diminish appreciably the output of the screws, as here the drag flow acts in the same direction. If we suppose, as we have done before, that the material within the channels "slips" on the barrel surface, because of the drag action of the screws it must move through these passages in a spiral around the two screws or around the "equivalent screw" (Fig. 2-19).

Returning now to the concept of "equivalent screw," we can calculate the output of the screws in these machines. If the screws rotate at n rps, we think the equivalent screw as rotating at an equivalent rpm, which gives the same circumferential speed or:

$$\pi D n = \pi D_e\, n_e \quad \text{or} \quad n_e = n \frac{D}{2D - 0.9\sqrt{Dh}},$$

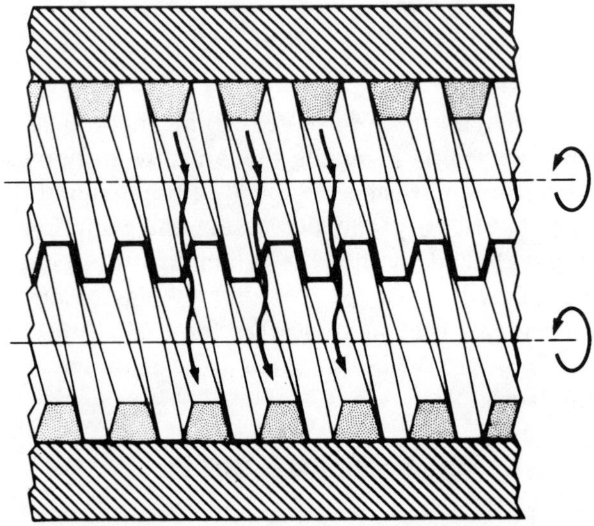

Fig. 2-19. Movement of the material around corotating twin screw.

With the help of the equivalent screw, the volume of the material that can be transported by the screws, or the output capacity, can now be written as the area of the channel of one screw multiplied by the length of the figure-eight-shaped channel around the two screws multiplied by the rpm of the equivalent screw:

$$Q_c = A \times S_e \times n_e$$

$$Q_c = \frac{1}{2}\pi Dh \sin\phi \cdot \frac{\pi}{\cos\phi}(2D - 0.9\sqrt{Dh}) \cdot n \frac{D}{2D - 0.9\sqrt{Dh}}$$

$$Q_c = \frac{1}{2}\pi^2 D^2 nh \tan\phi$$

This is the output capacity of the screws when rotated at n *rps.* From the Q_c, we must deduct the backflow q through all the passages and mechanical clearances. This depends, of course, on the size and shape of these passages, which are established when designing the screws, but also depends on the head pressure and the viscosity of the material. Tests have shown that when the screw's flight tip is equal to E, on the average, q varies between 10% and 15% of the calculated Q_c, or, in other words, the pumping efficiency is between .85 and .90.

The maximum output capacity of the screws can be written as for counterrotating machines, $Q_{max} = Q_c - q$, but here the pumping is much more efficient.

For a stated screw, Q_c is a function of the screw's speed n, which is multiplied by a factor χ, depending only on the design of the screws. This factor χ is a constant characteristic of the screws. In order to have the output in weight (kg/hr), we should multiply the result by the density d of the material. We have, then:

$$Q_{max} = \chi dn$$

where

$$\chi = \frac{1}{2} \pi^2 D^2 h \tan \phi$$

and because $\tan \phi = p / \pi D$, we can write:

$$\chi = \frac{\pi}{2} Dhp \text{ (cm}^3\text{)}$$

This constant is very useful in calculating Q_{max} and various other parameters at different rpm.

If we express all dimensions in centimeters, the melt density in gm/cm^3 and n in rpm, we have:

$$Q_{max} = \chi \text{ (cm}^3\text{)} \cdot d \text{ (gm/cm}^3\text{)} \cdot \frac{\text{rpm}}{60} = 0.06\, \chi \cdot d \cdot \text{rpm} \quad \text{(kg/hr)}$$

or to express it in pounds per hour:

$$Q_{max} = 0.13\, \chi\, d \text{ rpm} \quad \text{(lb/hr)}$$

For example, a twin-screw machine with intermeshing screws and conjugated profiles, having 6.6 cm interaxis and the following data:

$$I = 6.6 \text{ cm}; D = 8.2 \text{ cm}; h = 1.6; p = 3.14; \phi = 6°57'$$

rotating at a maximum speed of 34 rpm, according to the previous formulas and taking into account the pumping efficiency, has a maximum output between 73 and 104 kg/hr when counterrotating and between 112 and 119 kg/hr when corotating.

All these calculations about output only refer to the maximum possible output or output capacity of the screws for the two types of counterrotating and corotating screws. Some twin-screw machines are directly fed, and the amount of material is controlled by the screw's speed. In this case, the screws are full, as in single-screw extruders, and the output is directly related to the rpm and can be calculated with the previous formulas. This, however, links the output with the shear action of the screws, thereby limiting the flexibility of the machine and its ability to extrude heat-sensitive polymers at a high rate.

Many twin-screw extruders instead have a separate mechanism for controlling the amount of material fed into the screws. The actual output is then really given by the amount fed by this mechanism and only limited by the maximum extrusion capacity of the screws. Therefore, the above calculations, in this case, give only the possible output and not the actual output, which is only equal to the amount actually fed. As long as Q_{max} is larger than the required output, we can be satisfied that we have the machine we need.

Varying the amount fed means varying the output, and the feed can be increased up to the capacity of the screws. If more material than that is fed, the screws will refuse it.

In the above formulas, all parameters are fixed by the screw's design except the screw speed n and the viscosity, which affects q and is partially a function of the screw's speed. By increasing n within the limits imposed by other considerations such as type of material, type of extrusion, shear required, etc., both Q_c and q increase, and the output capacity can be somewhat increased, even if not linearly with n.

In these machines, the real actual output Q is therefore only a sizable fraction of the extrusion capacity Q_c and can vary between zero, when no material is fed, to close to one when the amount fed is close to the extrusion capacity at the rpm at which the machine is running (Fig. 2-20). The ratio Q/Q_c indicates the amount of the screw's filling or "filling ratio."

Another point we have to mention between counterrotating and corotating machines in regard to output is the difference in their operating speed.

Counterrotating screws cannot rotate at high speed because of the tendency of the screws to separate from one another and to press against the barrel, causing high wear.

Furthermore, high screw speed would increase the shear stresses near the barrel, already high, to a too high level.

Therefore, in forecasting the output capacity in real operation, it has to be taken into account that corotating screw machines can run at speeds up to three times that of counterrotating systems.

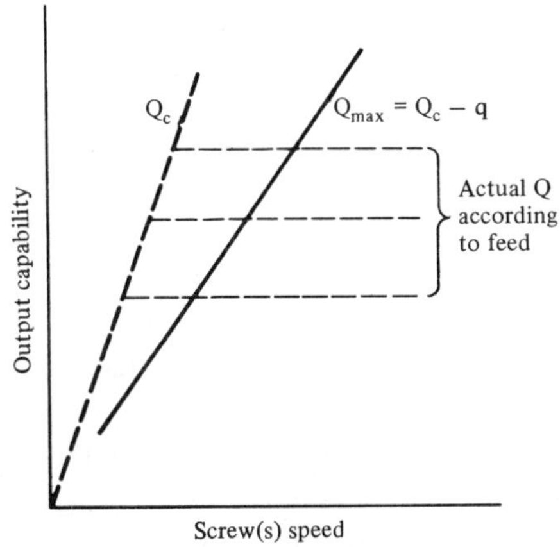

Fig. 2-20. Variation of output with screws speed in extruders with and without separate feed.

CONFIGURATION OF SCREWS

From what we have seen previously, the various parameters defining the screws can vary along the screws according to the function of the screws at each point.

Once the basic values of interaxis and diameters have been chosen, the screws can be then composed of various sections, each made with different pitch and/or different flight width so as to perform best the function we want them to perform within that section. These functions are:

Feeding
Melting
Mixing
Venting
Pumping

Practically every section, although made for a particular function, also partially performs other functions. So all sections, more or less, by the very nature of intermeshing twin screws, propel the material forward; a feeding section not only propels the material away from the feed opening but also starts the mixing; a melting section also mixes, and so does a pumping section, although in much minor proportion.

To enhance the functioning of some of these sections by retaining the material in them, special sections have been devised that have either very little or no propulsive action at all; among them are sections with interrupted flights, nonflighted, that is, cylindrical, sections, kneting discs, or counterflighted sections. Each section has a length appropriate to its function. In counterrotating screws, the total length may be divided into two or three sections, or it may be in one single part with changes in pitch and/or flight width defining the various sections. In corotating screws, the total length of the screws is usually divided in sections of about 3–4 diameters in length, but some sections can reach 6–7 diameters. The total length of the screws in both cases, because of the relatively small pitch, involves somewhere between 30 and 50 flights. The total length of the screws is determined by the sum of the length of the individual sections.

The ratio length to diameter (L/D) has no significant value in twin-screw extruders, and it usually varies from 8 to 15, although in modern machines, as more sections are added for perfecting the work of the screws, it now reaches 20–25 L/D. These different sections do not have to be physical sections of the screws at all but can be defined along the screws by changes in flight shape, width, or pitch, or any combination of the three. Cuts in the flights or any other special arrangement, like nonflighted, counterflighted part or kneting discs group, may define a section.

The screws can then be made, in one piece or in parts, with as many sections as needed, in proper order so that the screws as a whole can perform all the jobs required for extrusion. Because of their great differences in *modus operandi*, the screw's configuration of counterrotating and corotating screw extruders are also greatly different. We will therefore examine them separately.

3
Counterrotating Extruders

The main point differentiating the various sections of the screws in counterrotating extruders is conjugation. As we have seen, whatever the shape of the flights, if they are perfectly conjugated in a certain section of the screws, that section will have a positive pumping ability but none or very little mixing. On the contrary, if they are not conjugated, mixing increases with loss of positive pumping.

By a judicious choice, a screw configuration can be achieved that performs as desired.

Once the interaxis I and the screw diameters D and D_i have been established, there are three basic approaches to the design of counterrotating screw configuration: constant pitch with variable flights width, constant width of the flights with variable pitch, or, in some cases, varying both parameters. Sometimes a special screw configuration has been adopted involving multiple start threading of the screws.

The earliest screw configuration consisted in one section only with constant pitch and flight width, but the need for better performance required differentiation among the various sections in order to make each one best perform its job.

Soon two sections were developed, a feeding and mixing section and a pumping section, the two sections either maintained the same pitch and changed the flight width or reduced the pitch, maintaining the conjugation.

Later yet, multiple section screws were designed for feeding, melting and mixing, venting, and pumping.

Each manufacturer evolved its own way to give the screws the right configuration so that they would do the proper job at the proper point. Some preferred to maintain the thread continuous along the screws, and others divided the screws into variously threaded parts or sections.

Trudex, Maprē, and Fellow maintained the same pitch along the whole screw; Trudex cut the threads to achieve mixing; Maprē kept the same pitch and channel width and increased the flight width continuously from rear to front, thereby slowly transforming a mixing section with large passages and little shear into a melting and, further, into a pumping section with conjugated screws; Fellow reduced the flight but not the channel width at the middle of the barrel only and created a screw with a feeding and slightly mixing, a mixing, and a pumping section.

Others, like Anger, acted on the pitch, actually dividing the screws in parts, with one part a feeding-mixing section and a mixing-pumping section (with cuts for enhancing the mixing effect) and another part of smaller pitch and conjugated screws for pumping. The MOI Company used screws with one part with well-conjugated screws for feeding, another of smaller pitch and narrower flights as mixing-melting, and a third with the same pitch but conjugated screws as pumping section.

Pasquetti-Schloemann instead kept the screws conjugated all along and divided the screws in three sections: a feeding section with a screw with long pitch and three starts, the second with shorter pitch and two starts, and the pumping section with one start and an even shorter pitch.

In the case of conical screws (Cincinnati-Milacron), the screws are in various sections but in one piece, with no possibility of exchanging place between sections because of the continuous variation of screw diameter and interaxis.

An interesting solution is that of Kesterman, whose screws, starting with a long pitch and conjugated screws, have a wide flight into which, starting about midbarrel, another channel is cut so as to create a two-start screw. This type of screw, although maintaining conjugation all along, in the transition zone thoroughly and positively mixes the various flows and then, with a two-start screw, well conjugated, positively pumps it out of the barrel.

FEEDING

As we have seen, some counterrotating extruders have a separate mechanism to control the amount of material fed into the extruder, while some others use the positive displacement action of the screws to measure the material fed. In this latter case, a good propulsive ability is required of this first section of the screws; therefore, the screws have to have a good conjugation.

This can be achieved by having rectangular flights (which can be used in counterotating screws) of the same width of the channels with conjugated profiles.

If a feeding mechanism is used, although some propulsive ability is required to take the material away from the feed opening and into the barrel, there is a requirement for mixing and equalizing the feed rate. For that reason, many feed sections have reduced flight width in this section so as to increase the mixing action. The material fed into the barrel has a certain bulk density that is lower than the density of the melted material, as many voids are left between the particles. According to the size and shape of the particles of material fed, the material will occupy a larger volume than the same weight of melt. The feeding section of the screws must therefore have a larger volumetric capacity than the sections further downstream.

Fig. 3-1. Counterrotating twin-screw extruder (Toshiba).

Fig. 3-2. Counterrotating twin-screw extruder (Reifenhauser Nabco).

While being pushed forward, the material is compressed, or the amount of voids between its particles diminishes and the volume is reduced by the action of the heat and of the screws.

We must therefore have at the feed zone a section of screws with a much larger volume of channels to accommodate the light bulk density material corresponding to the maximum extrusion capacity of the machine calculated in weight.

In the feed section of counterrotating screws, the direction of rotation assumes great importance; in order to feed as much material as possible, in most types, the screws rotate outward at the top and inward at the bottom. This way, the material can fill the channels instead of being squeezed between the screws where there is little free volume available for material intake (Fig. 3-4).

As conjugated profiles would contain the material in closed chambers, thus limiting the compression of the material and the

Fig. 3-3. Counterrotating twin-screw extruder (Schloemann).

escape of air, this section is usually intentionally made nonconjugated with large clearances. This, however, also limits the propulsive ability needed in this zone. Subsequently, a second section shall have smaller channels in order to be filled with the "compressed" or less bulky material.

MELTING

To melt a material is to change its state by reducing its viscosity. For some materials, such change is rather sudden; for most thermoplastics, it is a gradual transforming of the material from solid, first, to a

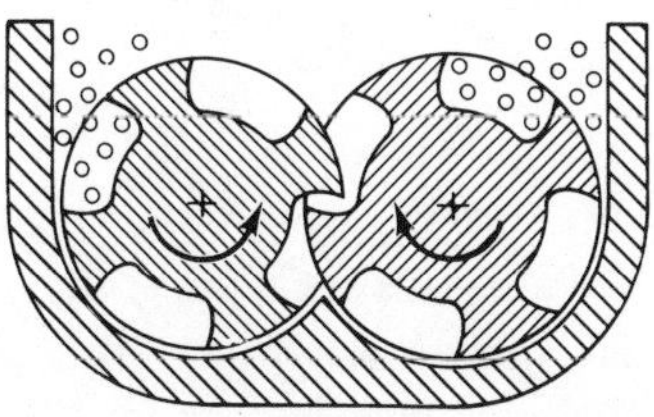

Fig. 3-4. Feed intake of counterrotating screws.

puttylike or high-viscosity mass and then into a more watery low-viscosity fluid.

All thermoplastics, non-Newtonian materials, are affected by temperature and/or shear increase. As shown in Figure 3-5, at a stated temperature, the viscosity drops very sharply with a shear rate increase. In any extruder, it is the combined work of heat and shear that melts the material. Shear rate within the channels can be calculated with the well-known formula:

$$\dot{\gamma} = \frac{\pi D n}{h}$$

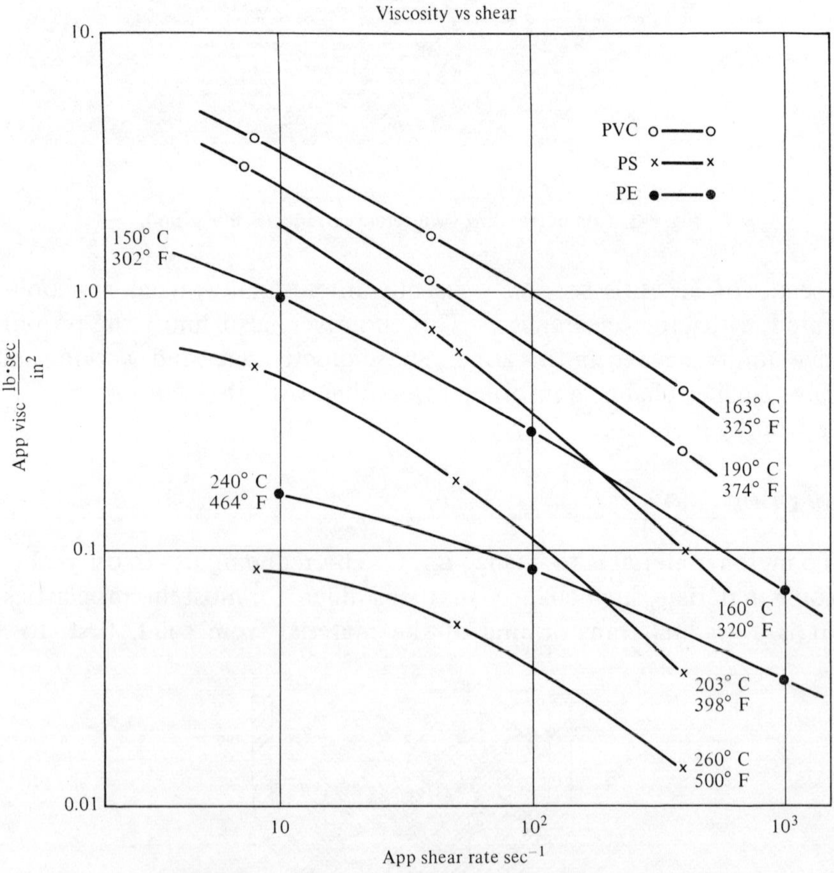

Fig. 3-5. Viscosity vs. Shear for various materials.

All the sections of the screws where the plastic is still unmelted have a channel volume much larger than the channel volume of the pumping section, as the bulk density of the solid material is always lower than that of the melt. In those sections, the degree of filling is therefore much smaller than one. The length of time in which the material resides in these sections is given by (see section on Residence Time):

$$T = \frac{Np}{V_e}$$

During that time, the material is still in a solid state since when the filling degree is low, shear is also very low, and so the viscosity remains high.

Even in a pumping section, where the shear is higher than in other sections, shear alone would not be sufficient to melt the material unless it had already reached a certain temperature. The necessary heat to reach that temperature is given to the material through the barrel by heater bands placed around the barrel itself.

According to the average specific heat of the material before the melting point and its starting temperature, a certain amount of BTUs are required per each pound of material (Fig. 3-6; Table 3-1) or:

Heat required = specific heat C_p X output X temperature differential

and this heat has to be furnished during the residence time T.

As the heaters must be able to supply the needed heat when the output is the maximum possible, that is, when the output is equal to the output capacity of the pumping section, we insert the value of Q_c. We have, then, the power required of the heater bands, which, apart from the losses, is:

$$\text{Watts} = C_p \; \Delta t \, \frac{\pi^3 D^3 \tan \phi_p \tan \phi_F}{L} \; n^2$$

where ϕ_p is the pitch angle in the pumping section, ϕ_F the pitch angle in the preceding section, and n the maximum rpm of the screws.

More practically and simply, as the filling ratio is never equal to one and the actual output is not proportional to the screw's speed, we can take a different approach, using the maximum actual output.

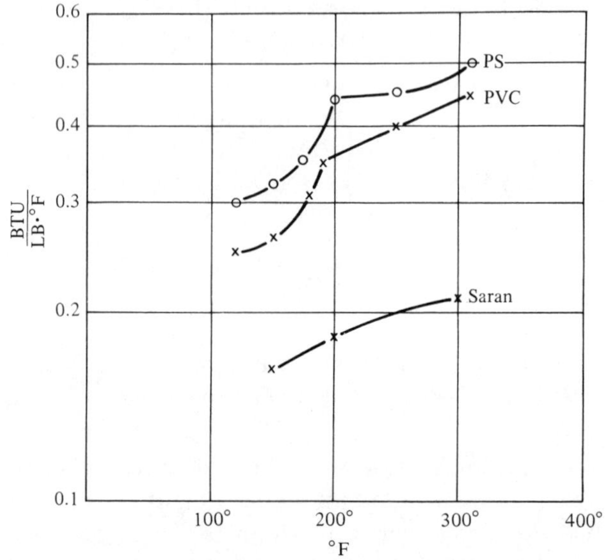

Fig. 3-6. Variation of specific heat vs. temperature (sharp point denotes melting.)

Inserting all needed coefficients to express all terms uniformly, we can write:

$$kW = C_p \left(\frac{\text{Cal}}{\text{kg} \times {}^\circ\text{C}}\right) \times 1.16 \times 10^{-3} \left(\frac{\text{kWh}}{\text{Cal}}\right) \times \Delta t\ ({}^\circ\text{C}) \times Q \left(\frac{\text{kg}}{\text{hr}}\right)$$

$$kW = 1.16 \times 10^{-3} \times C_p \times \Delta t \times Q$$

Table 3-1. Specific Heat of Various Materials at Constant Pressure.

Material	Specific Heat C_p
ABS	0.3 - 0.4
Acrylic	0.35 - 0.6
Nylon 6/6	0.4 - 0.63
Nylon 11	0.42 - 0.58
Polycarbonate	0.28 - 0.3
Polyethylene LDPE	0.5 - 0.55
Polyethylene HDPE	0.55 - 0.6
Polypropylene	0.45 - 0.5
Polyethylene UHDPE	0.7 - 0.78
Polystyrene	0.32 - 0.35
PVC rigid	0.2 - 0.28
PVC flexible	0.3 - 0.5

which, if we, as first approximation, assume a Δt of 170°C and an average specific heat of .4 Cal/kg × °C (≈ Btu/lb × °F) within that temperature range, we have:

$$1.16 \times 10^{-3} \times 0.4 \times 170 = 0.078 \text{ kWh/kg} \approx 0.1 \text{ hp/lb} \cdot \text{hr}$$

We can then design the appropriate heater bands.

If the machine is not well insulated thermically, we will have losses through irradiation, conduction, and convection. These usually amount to quite a large proportion of the heat needed and may require a doubling of the power of the heater bands. On the other hand, the energy transferred to the material at the melting point through shear will increase its temperature so that this can somewhat compensate for the losses (See section on Efficiency).

MIXING

One of the most important functions of an extruder is a good mixing action. While the polymer moves through the barrel, every particle may undergo different treatments. Depending on the specific path followed by each one of them, shear, temperature, and residence time can vary widely and influence the viscosity and the final characteristics of each particle of material. In order for the entire mass of material to have the same characteristics, each particle has to exchange position and mix with others during the extrusion process so that the average treatment is the same for all of them.

Mixing can be basically divided into two actions: disintegrating and homogenizing. The first consists of disaggregation of particles and dissolution of gels; the second, a redistribution of particles uniformly throughout the bulk.

To have thorough mixing, we must have not only movement of the material within the channels but also a partial transfer of material from channel to channel. This type of mixing will not only dampen the feed inconstancy but also distribute additives, such as colors, lubricants, stabilizers, or nucleators, as the case may be, and render the mixture uniform. This has to be accomplished in a reasonably short length of the barrel and has to be done without developing too much shear, as otherwise the material is likely to become overheated.

In counterrotating twin-screw extruders, the high shear to which the material is subjected where the screws are conjugated handles the disintegration; however, in these sections, there is no connection between channels, so that homogenization of the material cannot be done within them. Mixing sections have to be introduced that have the flights thinner than the channels are wide (nonconjugated) or where, with whatever other means, passages are left between the flights. These mixing sections have nonconjugated profiles, and although the high-speed flow of material tends to keep these passages free, there is no positive self-cleaning action of the screws themselves, leaving the possibility of some material hangup with consequent potential degradation.

The solution adopted by Pasquetti and Kesterman of changing the pitch and the number of thread starts, either in section or gradually, keep the screws conjugated at all times, thereby eliminating this potential problem.

Another solution to the mixing problem (Trudex, Anger) is the use of interrupted flights. Here, in a basically pumping section with conjugated profiles, cuts are made in the flights; because of the pressure gradient within a flight due to the counter direction of rotation, the material passes through these cuts from one channel to the other, thereby being homogenized.

The mixing action of counterrotating screws, although a positive action in the sense that it is not dependent on extrusion conditions, cannot be easily calculated, as we do not know how much or what part of the material is transferred from one to another channel.

VENTING

In some cases, it is necessary to vent or extract from the melt the gases trapped or contained in it.

A hole made in the barrel at the proper position will do just that. However, the material at that point must not be under pressure; otherwise, it will also come out of the barrel. In twin-screw extruders, either counterrotating or corotating, because of their positive pumping action, this can be easily accomplished by having at that point a section of screws with much larger output capacity than the previous pumping section so as to greatly diminish at that point the degree of filling of the screws. These sections must be followed by another pumping section that positively will take away the material from the venting area and propel it further. The venting part of the barrel will therefore be at a very low degree of filling (almost empty), and the material will be under no pressure.

In this venting zone, we also need to mix the material in order to bring every particle of it to the surface where the gases dissolved or trapped in it can come out easily. This can be accomplished by choosing for this section a mixing configuration with interrupted or nonconjugated flights.

PUMPING

Extrusion is by definition the operation of forcibly pushing a melted material through a die. In order to do that, the material has to be under pressure, and the pressure has somehow to be created by the extruder's screws.

Unlike single-screw extruders, the building up of the required pressure in twin-screw extruders is a positive action that is almost independent of operational conditions. Actually, counterrotating screws work like a gear pump, the latter being an extreme case of intermeshing screws with pitch angle $\phi = 90^\circ$ and where, therefore, the screws become gears.

When the intermeshing screws are conjugated, separate volumes of material remain sealed between the screw threads and by the rotation of the screws are transported axially and forced out of the front of the barrel. Backflow is negligible and practically not affected by the pressure at the die.

The counterrotating flow of material, however, and the consequent pressure differential around the channels pushes some of the material through the mechanical clearances at the plane of the screw's axis into a different channel. This accounts not only for a localized heating due to high shear but also a small loss of pumping ability. The passage of material through these clearances compares to a sort of internal extrusion; the resistance offered by these clearances to the flow depend on the cross-section of the openings, the speed of extrusion, and a coefficient related to the shape of the openings.

For example, in a dic, a given cross-section of 1 sq in shaped in a rectangular section .8 × 1.25″ has certainly less resistance than the same cross-section with dimensions .04″ × 24″. Assuming the smallest dimension as the thickness B, the width of the opening A, and the length L, we can say that this resistance is (Bernhardt):

$$R = \frac{12L}{A \times B^3}$$

The inverse of the resistance $1/R$ can be called conductance K. Through the slits left by the mechanical tolerances, according to the Hagen-Poiseuille equation, passes a flow:

$$q = \frac{PK}{\mu}$$

and as the viscosity μ is locally very low due to the locally high shear, the amount q assumes a certain value that detracts from the pumping ability of the screws.

The higher the rpm, the higher the output capability, but also the higher is the drag flow and therefore the pressure differential within a channel; this, together with higher shear and corresponding lower viscosity, may increase the value of q to the extent of limiting the output increase expected when increasing the rpm. The output curve, as a function of the rpm, flattens out, limiting the advantage of higher screw speed. Furthermore, the repeated shear action through these small clearances may increase the stock temperature well above the material's limit.

Theoretically, one flight only may suffice to bring the pressure up to the value necessary for extrusion. However, because of leakage flow around the screws and because of some backflow through the slits between the screws, more than one flight is needed. In any case, the pumping efficiency of this type of screw is so great that usually no more than two or three flights are involved in the pressure buildup.

A reduction in the flight width or a loss of conjugation caused by any other change will drastically decrease the pumping ability of counterrotating screws.

RESIDENCE TIME

As long as the screws have a pumping configuration, that is, in counterrotating screws when they are conjugated, the material being contained in closed chambers must advance one pitch per each revolution of the screws. The time required for the material to advance for a length $L = Np$ is:

$$T = \frac{Np}{V_e} = \frac{N\pi D \tan \phi}{n\pi D \tan \phi} = \frac{N}{n}$$

where V_e is the axial speed of one pitch per revolution.

If the whole screw had a pumping configuration, the residence time would be very short, and not enough time would be allowed for heat transfer, proper melting, and proper mixing. For this reason, at a certain point along the barrel, the screws are made so as to lose conjugation, or as we have seen previously under mixing, their flights have cuts to permit shuffling. As the screws have so lost most of their propulsive ability, the material will fill completely the screws and move back and forth between flights. This makes the residence time a little longer than the one calculated above, although not too much.

High screw speed will, of course, reduce residence time to the extent that no sizable amount of heat can be transferred to the material through the barrel from outside sources, but at the same time the increase in shear will become sufficient to melt the material. Sometimes, however, with high speed, the shear may become exceedingly high. Low screw speed, on the contrary, reduces shear, which, in most cases, is desirable, increasing at the same time the length of residence, thus permitting better heating of the material through the barrel. Generally speaking, this latter solution is the one most often adopted by the designer of this type of extruder.

POWER CONSUMPTION

All these operations of mixing, melting, and pumping require and use up some power. Power is delivered by the main motor through a gear box to the screws. From this point of view, the work of the screws on the material can be divided according to the place where it is performed. Very little power is used for rotating the screws as long as there is no material in the barrel; with material in it, power is needed.

Although in all twin-screw extruders, with their deep channels and slow screw speed, the shear within the channels is much lower than in single-screw extruders, we will first calculate the power dissipated as shear in the channels of the filled flights. In first approximation, this is given by:

$$Z = \mu \dot{\gamma}^2 V$$

where V is the volume involved. This is equal to the area of the channel by the length of it around the two screws less the intermeshing parts, that is, by the length of the equivalent screw. The volume is therefore:

$$A \times S_e = \frac{1}{2}\pi Dh \sin\phi \; \frac{\pi D_e}{\cos\phi} = \frac{\pi^2 DD_e h \tan\phi}{2}$$

and the power dissipated is:

$$Z = \mu \; \dot{\gamma}^2 V = \frac{\bar{\mu}\, \pi^4\, D^3\, D_e \tan\phi}{2\,h}\; n^2$$

which should be multiplied by the number of interested or filled flights N.

In counterrotating screws, N is usually the total number of flights with the exception of those that are cut and those of the venting zone.

If all values are given in centimeters-grams-seconds (c.g.s.), the calculated value has to be multiplied by a transformation factor of 1.3142×10^{-7} in order to obtain the result in horsepower.

Besides within the channels, other points where shear develops are (1) between the flights tips and the barrel, (2) between the flight tips of one screw and the screw root or the bottom of the channel of the other screw, and (3) between the flanks of the flights in the intermeshing region.

These actions can be defined by equations that derive from the power equation (Schenkel), the roller mill equation of Ardichvili, and others. Before we can calculate the amount of shear in these regions, we should remember that in counterrotating screws the pressure differential around the screws within the channels separates the screws, pushing them against the barrel. This may change the results, as at a point around the circumference we may have metal-to-metal contact.

Assuming ρ as the clearance between the screws and the barrel, we have:

$$Z_c = \frac{\mu v^2 A}{\rho}$$

where A is the effective area of the working surfaces where the shear is developed, and v their relative speed.

Recalling now the definition of equivalent circumference as the sum of the circumferences of both screws less the two parts overlapping in the intermeshing region and assuming e to be the average flight-tip width and substituting for one flight, we have:

$$Z_c = \frac{\mu\, v^2 e C_e}{\rho} = \frac{\pi^2 D^2\ e C_e}{\rho}\quad \mu n^2$$

For the shear in the mill created by the tip of one screw and the bottom of the channel of the other, we have first to see how wide the flight is in relation to the channel bottom. If the flight is thin and the ratio between the two is small, this term has no importance, as practically very little shear is developed under this condition. Only if the screws are conjugated, that is, when the flight tip is as wide as the channel bottom, will the roller mill shear assume some meaningful, albeit small, value. The speed of the two surfaces does not differ much, as they move in the same direction; the only difference is due to their different diameters. Their relative speed is then:

$$v = \pi Dn - \pi Dn\ (D - 2h) = 2\pi nh$$

The basic formula to use is:

$$Z_s = \mu v^2\ (D - h) \left(\frac{1}{\epsilon} - \frac{1}{\omega}\right)\ l$$

As the thickness ω of the bank of material behind this slot is much bigger than the clearance ϵ, we can disregard it, and as two couples of flight tips and channel bottoms of width e are involved in each turn of these sections, we have:

$$Z_s = \mu v^2\ \frac{D - h}{\epsilon}\ 2e$$

Substituting and recalling that $D - h$ is equal to the interaxis I, we can write:

$$Z_s = \frac{8\pi^2 h^2 eI}{\epsilon}\ \mu\ n^2$$

We have then to examine, even if summarily, the third point, mentioned above, which is the passages between the flight's flanks. Here, the flight's width, and conversely the clearance left, is of great importance. When the screws rotate, the material within the channels, rotating with them, tends to move and accumulate on one side of the intermeshing region at a speed corresponding to the screw's speed and pass through the slits between the screws. These slits, either due to the need of mechanical clearance or purposely left between the screws threads, have a width equal to the depth of the two flanks of the flights $2h$ and a thickness of σ chosen by design. Through these slits, the material is somewhat forcibly extruded into the next channel. The amount passing through this sort of die is q_i, which is equal to the amount propelled by the drag flow in the channel less the amount q_c "rejected" as pressure flow:

$$q_i = Q - q_c \quad \text{and} \quad Q - q_c - q_i = 0$$

The pressure flow, as the definition implies, and the amount flowing through the slits are both proportional to the pressure created and to the relative dimensions of the channels and the slits. Recalling the definition of conductance, we have:

$$q_c = \frac{PK_c}{\mu} \quad \text{and} \quad q_i = \frac{PK_i}{\mu}$$

As the pressure is the same, substituting and solving for P:

$$P = \frac{Q\,\mu}{K_c + K_i} \quad \text{and} \quad q_i = \frac{QK_i}{K_c + K_i}$$

which is independent of pressure and of the material viscosity and depends only on the ratio of the slits to channel dimension. In counterrotating screws, the flights in the intermeshing region move in the same direction, and the only relative motion is because opposite points on the two surfaces rotate on different diameters. The maximum speed differential is at a point located at the tip of one screw, which rotates on a diameter D, while the opposite point of the other screw rotates on a diameter D_i. We have, then:

$$v_2 - v_1 = \pi Dn - \pi(D - h)n = \pi nh$$

The speed differential between the two surfaces diminishes to a line where the diameters of rotation are equal, to increase again to the same value, although in opposite direction. The average speed is $v = \frac{1}{2}\,\pi\,hn$. The surfaces involved have a shape that, due to the intermeshing of the two screws, is assimilable to two triangles, each of height $\frac{1}{2}\sqrt{D^2 - I^2}$ and base h, joined by the bases (Fig. 2-13). Therefore, the power dissipated in each filled flight amounts to:

$$Z_\omega = \frac{\mu v^2 A}{\sigma} = \frac{\mu v^2 h \sqrt{D^2 - I^2}}{2\sigma} = \frac{\mu \pi^2 h^3 \sqrt{D^2 - I^2}}{8\sigma} n^2$$

Calculation of Z_ω shows that the power dissipated into the material at this point is very limited due to the concurring motion of the screws in the intermeshing region.

The amount of power required increases with a decrease of the size σ of the gap, but, more importantly, the amount of material passing through these gaps greatly decreases with a decrease of σ. Decreasing the gaps reduces the total amount of power but markedly increases the amount dissipated per unit of volume of material subjected to it. When σ is very small, the relatively large quantity of heat thus transferred to a small quantity of material may increase its temperature to above the permissible limit and make it degrade or decompose (which will then show as black spots in the extrudate) without any noticeable increase of the total power absorbed from the main motor. For a given extruder rotating at a fixed rpm, the temperature increase of the material passing through these gaps between the screw flanks is shown in Table 3-2 as a function of the gap size σ, and using: $Q = 34$ cm^3/sec; $K_i = 0.013 \times \sigma^3$; $K_c = 0.066$; $Z_w = 447/\sigma$; $n = 20$ rpm.
Adding to these equations the power needed for pressure buildup at the die, which is

$$Z_p = \frac{Q^2 \mu}{K_f}$$

we have the total power consumption within the screws:

$$Z_{\mathrm{TOT}} = N(Z + Z_c + Z_s + Z_\omega) + Z_p$$

$$Z_{\mathrm{TOT}} = \left[\frac{\pi^4 D^3 D_e \tan\phi}{2h} + \frac{\pi^2 D^2 eCe}{\rho} + \frac{8\pi^2 h^2 eI}{\epsilon} + \frac{\pi^2 h^3 \sqrt{D^2 - I^2}}{8\sigma}\right] \bar{\mu} N n^2 + \frac{\mu Q^2}{K_f}.$$

This equation, from the design point of view, tells us immediately that the required power, or size of the required motor, increases exponentially with the physical size of the extruder and poses a limit

Table 3-2. Increase of Temperature of the Particles Going Through a Gap of Size σ.

σ	q_i	Z_ω	Z_ω/q_i	Δt
0.4	4.38×10^{-1}	0.14×10^{-3}	0.6	2
0.2	5.50×10^{-2}	0.29×10^{-3}	9.3	31
0.1	6.97×10^{-3}	0.58×10^{-3}	148	493
cm	cm^3/sec	hp	Cal/Kg	°C

to its size. We can also see that once the size of the screws has been established, the factors affecting the power consumption are:

$$Z_{\text{TOT}} = f\left(\frac{\mu Q^2}{K_f};\ \bar{\mu};N;n^2\right)$$

As the final viscosity depends inversely on the screw's rpm, the net result is that a change in the screw's speed does not affect, except within a narrow range, the power required.

Apart from the screw's speed, for a given material and a given extruder, the actual output is the factor that basically affects the power dissipated by the screws.

4
Corotating Screws

In corotating twin-screw extruders, because of their particular geometry, various sections for different functions can be made and still always maintain the conjugation of the screws.

In these machines, each section has a much more definite job than in counterrotating screw extruders, so it can be made as a separate part and assembled where its particular job is required without affecting, or only in a minor way, the work of the other sections. This is the reason why some manufacturers of corotating screw extruders make screw sections for different functions and either assemble them as complete screws themselves (LMP/Colombo) or furnish them to the customers for them to assemble as needed (Werner & Pfleiderer). Other manufacturers instead prefer to make the screws in one piece, or two, with the various sections cut on them according to their own design.

The simplest configuration of the screws requires only two sections: a feeding and a pumping section. Once the interaxis and the screw diameters are established in accordance with the intended production output, the first section will have a larger diameter and an $e < E$; the second, a smaller diameter and an $e = E$. Historically, these were the earliest screw configurations, and very small and very short extruders were manufactured; later, with increasing needs, more sections with specific tasks were added.

First another section was added between the two, with intermediate values of diameter and of e in order to handle the material, which became less bulky through heat and handling by the screws. This section had a higher screw-filling ratio than the feed section but less than the last or pumping section. Then a large-pitch feeding section, as shown in Figure 4-5 was also added to better feed very bulky material.

Fig. 4-1. An early model of twin-screw extruder (LMP/Colombo).

Finally, the screw configuration reached the sophistication of today, where as many as eight or nine sections compose the screws: (1) a feeding section always with large diameter and small *e;* (2) a "compression" section and a pumping section with smaller diameter, smaller pitch, and larger *e;* this pumping section extrudes through (3) a counterflighted section, which we will see better later, where the material melts and where a high-pressure zone avoids gas return; (4) a series of kneting disks or a mixing section with smaller diameter and pitch and average *e* to regularize the flow of material into the next; (5) a venting section with as large a diameter as possible, high pitch, and small *e;* and, finally, (6) one mixing section or kneting discs, as previously and (7) a pumping section with small diameter and a screw tip width equal *E.* In some cases, an extra mixing section is added to cool the extrudate.

Fig. 4-2. Corotating twin-screw extruder for compounding (Berstorff).

Figure 4-3. Corotating twin-screw extruder (Werner & Pfleiderer).

Fig. 4-4. Corotating twin-screw extruder (LMP/Colombo).

In order to mechanically extract the screws from the barrel, the diameter of the screws must be either constant or increasing from rear to front so that the screws can be slid to the front or decreasing from rear to front so that the barrel can be slid to the front and expose the screws. As the feeding sections require a large channel volume to handle the bulkier material fed, this latter solution is preferable. The screws are divided in parts of slightly decreasing diameters from rear, or feeding side, to the front, or extrusion, end of the barrel.

Once this has been established, the other variables at our disposal to change the configuration and therefore the function of the various sections are the pitch and the screw-tip width *e*.

FEEDING

Corotating screws convey the material into the channels and around the two screws (Fig. 4-6).

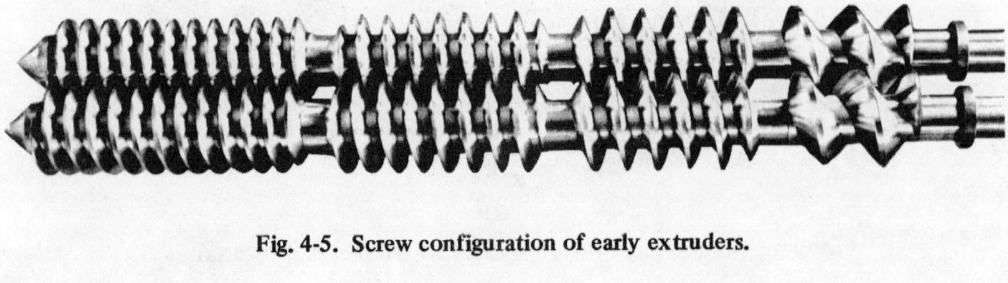

Fig. 4-5. Screw configuration of early extruders.

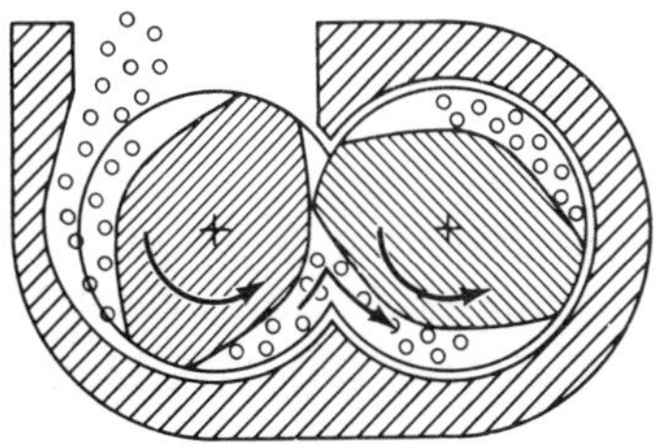

Fig. 4-6. Feed intake of corotating screws.

Every time the material reaches the intermeshing area, it is compressed, as it has to go through the wedge-shaped passages.

For good homogenization of the material fed, conjugated profiles with reduced tip width are usually adopted so that large volume intake, good compression, and thorough air escape together with good mixing, but still maintaining positive propulsion, is achieved.

As we have seen before, in most corotating twin-screw extruders, the output capacity of the screws has nothing to do with the actual output. This latter is determined by the amount of material actually fed into the screws by a separate mechanism measuring and dropping or delivering into the feed opening the amount to be extruded. As long as this is smaller than the output capacity, the screws will take it and extrude it. The extruder must therefore have a "filling ratio" less than one.

The reduction is bulk density due to compression of the particles of material and the difference in density between bulk and melt requires that the next section have a smaller volumetric capacity. Although the feed must be constant on the average, the material may be fed in a not absolutely constant way, but pulsating or, let us say, in handfuls. The feed section must therefore have a pitch and a flight-tip width so chosen that the volume of the channel is large and its ability of mixing is somewhat large so as to mix and equalize those possible pulsations while still propelling the material away from the feed opening.

The positive action of pumping, which is partially always present no matter how the flight-tip width is chosen, will propel any kind of material and forward it to the next section of the screws. This has been proved with slippery materials like wax and with rolling stock like the small styrene pearls or beads as they come out of the polymerization process. This is a great advantage of this type of machine, as it permits feeding into the extruder materials that could not otherwise be extruded and the use of highly lubricated formulations.

The volume of the channel needed in the feed section can be calculated knowing the melt density and the bulk density of the material to be used, or if many materials could be used, the worst condition of a material very bulky at feed and of light specific weight, as polyethylene regrind.

For example:

$$\frac{\text{melt density}}{\text{bulk density}} = 1.5\text{–}1.8$$

The ratio of the two represents the increase in channel volume that a feed screw section should have in relation to the pumping sections where the material is completely melted. In first approximation, as the volume of the channel per turn is:

$$V = A \times S = \frac{1}{2}\pi^2 D^2 h \tan\phi$$

either we can act only on the pitch, or we can enlarge the screw's diameter D of the feed section to the extent possible without reducing too much the screw's shaft diameter D_i (we must remember that $D + D_i = 2I =$ constant), at the same time increasing the pitch angle ϕ.

The following section does not need to have such a large channel, and we can have a slightly smaller diameter or a reduced pitch and a more propulsive configuration.

MELTING

In twin-screw extruders, each section of the screws is designed primarily to do a particular job: mixing, pumping, etc. As we have seen, each section has therefore different parameters of which ϕ and e are the most important.

We have also seen that until some pressure is built up, the screws are only transferring and mixing the material but not really working on it, as the flights are not filled. If the flights are not filled, the well-known formula for shear does not apply, and only a very limited amount of energy is transferred to the material through shear. Also, when the flights are not filled, the material does not contact the whole inside surface of the barrel, thus reducing the heat transferred to it from the outside heater bands. The material would be somewhat heated but not melted; melting will only occur in that part of the barrel in which pressure builds up and the flights are filled.

When the flights are filled, then we can calculate the shear, but we must remember that, in our case, since we have twin screws, we must use not the screw's diameter but the equivalent diameter. The shear rate will be:

$$\dot{\gamma} = \frac{\pi D_e n}{h} \; (\text{sec}^{-1})$$

and the related power:

$$Z = \mu \, \dot{\gamma} \, V \cdot 10^{-10} \; (\text{Kw})$$

where V is the volume of the material involved.

This power, transferred into the material, which is already heated by the heater bands, will reduce drastically the viscosity or melt the polymer.

Unlike single-screw extruders, in which the whole screw is filled with material, the volume involved is not the volume of the total length of screws but only the volume of the length L corresponding to the number of filled flights.

V = area channel X equivalent elix X number of filled flights

$$V = \frac{\pi D h \sin \phi}{2} \cdot \frac{\pi D_e}{\cos \phi} \cdot N$$

$$V = \frac{1}{2} \pi^2 h D D_e \tan \phi \, N$$

so that the power dissipated becomes:

$$Z = \mu \frac{\pi^2 D_e^2 n^2}{h^2} \cdot \frac{1}{2} \pi^2 h D D_e \tan \phi \, N$$

$$Z = \mu \frac{\pi^4 D D_e^3 \tan \phi \, N}{2h} n^2$$

This power, dissipated as heat into the material, increases its temperature proportionally to the square of the rpm.

We can calculate this temperature increase with the formula given by McKelvey and reported in Fischer as:

$$\Delta_t = \frac{1}{b}\, ln\, \frac{\mu \text{ inlet}}{\mu \text{ outlet}} \cdot$$

In corotating twin-screw extruders, shear is almost two orders of magnitude lower than in comparable single-screw extruders; the length of the filled screw is shorter, the channel is much deeper, and, more important, their screws rotate at much lower speed. Therefore, the power dissipated by shear in twin-screw extruders is much lower, and in absolute value, very small. Non-Newtonian materials vary their viscosity according to heat and shear (Fig. 4-7). While the screws have a low filling ratio, the heat is transferred to the material from the external heaters, and its viscosity diminishes, following closely the lines of very low constant shear. As soon as the channels are filled, the shear developed not only introduces some more heat but also by its action lowers the viscosity of the material, which is rapidly dropped to the required value for extrusion.

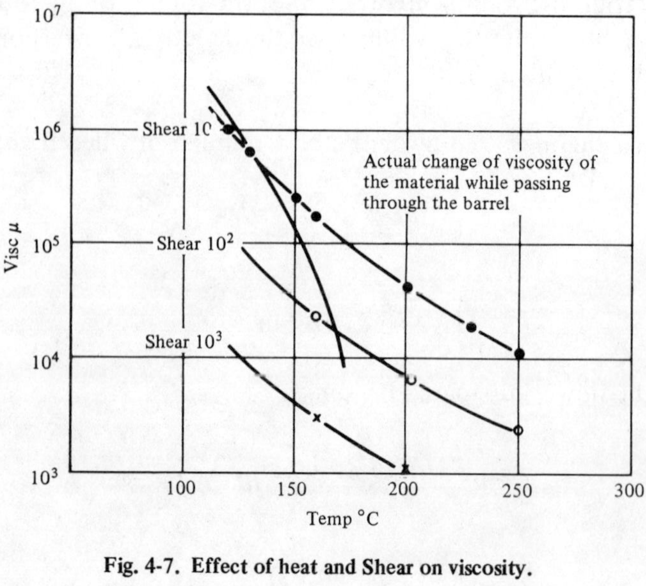

Fig. 4-7. Effect of heat and Shear on viscosity.

A typical heat balance for the melting section of a corotating twin-screw extruder shows that approximately 55% of the energy reaching the material comes from the screws and 45% from the heater bands.

In order to melt the material way back in the barrel, it is required that some pressure be built up at a certain point at midbarrel so that the flights can be filled up and shear as well as heat can be applied to it. To achieve that, some systems have been devised to "extrude" the material somewhere midbarrel.

One of the first arrangements tested was to simply eliminate the thread in a small part of the screws directly after a pumping section, reducing them to two rotating cylinders. This way, the screws at that point do not propel the material anymore, but the pumping section has to push it through the double anular passages, thereby compressing and somewhat extruding it.

In this system, however, the material underwent a high shear in those layers close to the rotating cylinders, with no mixing, and its various particles had different shear and heat history.

A second way tried was to cut passages in the flights so as to greatly reduce their propulsion ability, leaving the pumping to the section before them. This system kept the material very homogeneous, but the "extrusion" pressure reached was small; although this system can be useful for some easy-to-melt materials, it does not give good results with some others.

To achieve constantly good results with all types of material, another system has been designed. This system is based on a counterflighted section of the screws (Fig. 4-8).

If, after a pumping section, the following section of the screws has reverse pitch, the latter will try to refuse or pump backward the material pumped forward by the previous section. This way, where the

Fig. 4-8. Counterflighted slotted section (Colombo).

forward and backward pushed material accumulates, there will be a zone of high pressure, and many flights of the pumping section will be filled in order to create that pressure.

If cuts are made in the counterflighted section of the screws, the pressure will be relieved by making the material flow through the cuts, as through a pipe, and move forward. The higher the pressure required, the more flights will be filled and engaged in producing that pressure.

Every restriction existing in a flow creates a resistance to that flow; the smaller the passage, the larger the resistance, the thinner it is for a given area, the more again the resistance increases. As the resistance R_d is inversely proportional to AB^3, where A is the radial dimension of the slots and B its width, we can write:

$$N \equiv P = Q\mu R_d$$

The output Q for continuity must be equal to the one furnished by the pumping section; therefore, the pressure buildup in the joint between the sections is proportional to the viscosity and inversely proportional to the cube of the slot's width.

Assuming or calculating the viscosity, we can then predetermine the pressure and therefore the number of flights filled (length) of the pumping section by accordingly making the slots wider or smaller.

As soon as the material enters the "pipes" formed by the slots, however, it does not actually flow through them, but while it moves, it is also shuffled by the interrupted flights of the reverse flighted section so that homogeneity of the material is achieved and maintained. Within this section, the actual motion is alternatively a flow one pitch long through the slots and then a step backward for a part of the pitch pushed by the reverse flight. This greatly increases the shear to which the material is subjected right at the point where we want the material to melt.

As a side benefit, the high-pressure zone around the screws at the joint between direct and reverse flighted sections creates a seal that impedes the flow of gases backward toward the feed opening. With this device, the position along the barrel where melting is going to occur can be established.

The combination of various types of screw sections, with larger or smaller pitch and therefore volumetric capacity Q_c, large or smaller flight tip e, direct or reverse flights with different sizes of slots, etc., give these extruders maximum flexibility, so that we can melt the material where we want and reduce its viscosity only to the extent we want.

MIXING

In corotating twin-screw extruders with nonconjugated flights, crosswise paths are open to the material in the intermeshing area.

While the flights of one screw penetrates but does not occlude the channels of the other screw, it leaves ample passages all around. Material from the one channel not only follows the figure-eight path around the two screws but also passes through the voids left by the nonconjugating screws around each screw (Fig. 4-9).

The various paths that the material takes and the turbulence within the intermeshing area create a very good mixing.

In corotating twin-screw extruders with intermeshing conjugated screws, a mixing configuration of a certain section can be easily achieved, always maintaining the conjugation simply by acting on the flight-tip width.

When, while choosing *e,* we make it much smaller than *E,* as we have seen, each stream of material coming from a single channel of one screw is divided into two streams going into two channels, each stream joining with another coming from a different channel; the resulting mixture is then to be divided again at the next intermeshing point (Fig. 4-10). The ratio between the new passage created by a small *e* and the width of the total passage indicates the mixing ability of the screws.

If *e* is such that during the dividing action a significant part of the stream of material goes from one channel to the adjacent as well as to the next channel of the other screw, then the screws are predominantly mixing. On the other hand, if almost all the material goes to a single channel of the other screw, the mixing is minimal, as this would be a pumping configuration. The proportion of mixing versus

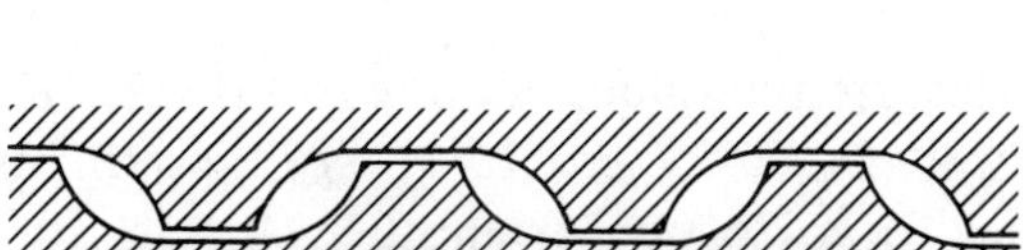

Fig. 4-9. Longitudinal section of corotating intermeshing nonconjugated screws.

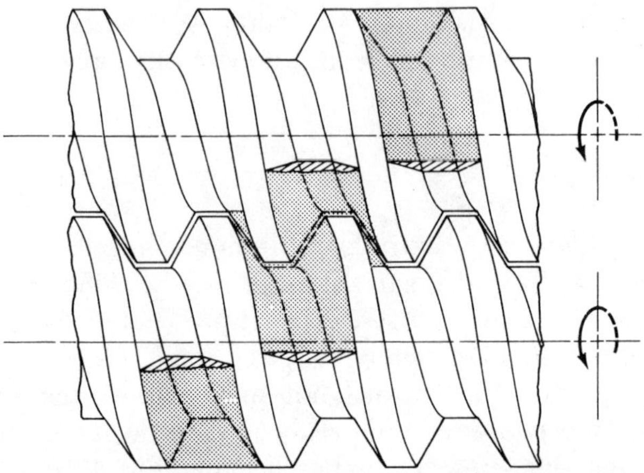

Fig. 4-10. Path of the material around the two screws with mixing configuration (flight tip width e<E).

pumping depends only on the choice of e in relation to E and can therefore be determined, according to need, when designing the screws.

The width of the two passages is determined by the choice of e. The width of the first is given by $E - e$, whereas the total opening is $p - 2e$; therefore, the mixing ability of the section of screws is:

$$m = \frac{E - e}{p - 2e}$$

For $e = E$ (maximum possible), the ratio will be equals zero, and for e equals zero (triangular flights), the ratio will be $m = E/p$, which, if we have chosen the maximum theoretically possible E, the ratio becomes $m = \frac{1}{2}$.

Although, theoretically, E can be between zero and $p/2$, as we have seen, for practical reasons, E is usually limited between $p/3$ and $p/4$. On the other hand, e must be smaller than E and comprised between zero and E.

If we now calculate the values of m, we see that for $E = 0$, triangular flights, m is always equal to zero; for $E = p/2$, no flights, the chances for the material to go to one or another channel are equal so that $m = \frac{1}{2}$. For the other two more reasonable values of $E = p/3$ and $p/4$, we have m varying between zero and .33 in the first case and between zero and .25 in the second.

A large E and a small as possible e will therefore give us a mixing section with a very good mixing capability. The greater is m, the greater the mixing ability of the section.

This division of the material into two streams occurs twice every turn of the material around its figure-eight path, or twice every turn of the equivalent screw. When the screw rotates at n rpm, the material moves around the equivalent screw at n_e revolutions. Usually, n_e is .62 or .65, but we can for simplicity take approximately 0.5 n; therefore, this division and joining of the material's streams occur 2×0.5 n, or one time per each revolution of the actual screws.

The successive division and merging accumulate, so that after n revolutions of the screws, there will be 2^n mixings. At 20 rpm, this means 2^{20} or 1,048,576 mixing per each minute of residence time. This kind of mixing spreads the irregularities of the feed and any uneven concentration of additives.

For example, if, during extrusion, a handful of different-colored material fills the first (feeding) flight for one revolution, this high concentration will be rapidly diluted and spread among many flights at a lower concentration in a short time (Table 4-1).

While this macrohomogenization takes place, the material has, of course, to go through the passages left by the screw's flights at the intermeshing regions. The passages formed by the combined flights have a shape different from the channels themselves. The channels are trapezoidal; these passages are basically triangular in form, and their area is defined by the angle of the flanks of the flights or, in other words, by the width of the flight tip (Fig. 4-11). This change of shape causes more intimate mixing of the particles, as the material in the channel has its core exposed for more uniform treatment.

Furthermore, the total area of the passages can be calculated, and knowing the area of the channel, we can see that the ratio of the two is always smaller than one and in the average is only .61 to .65. There is therefore a restriction to the flow of material at each intermeshing point or some sort of slight extrusion.

Due to the difference in the diameters in which the material lies, the relative linear velocities of the various layers is also different: greater for the layers near the tip of the flights and lower for those near the core of the screws. This causes a high turbulence, and the

Table 4-1. Distribution of the Material Within the Various Flights in a Mixing Section with m = 0.3 After Few Turns of the Screws.

Number of turns	1	2	3	4	5	6	7	8	9
0	0	0	0	0	0	0	0	0	0
0	100	0	0	0	0	0	0	0	0
1	30	70	0	0	0	0	0	0	0
2	9	4.2	4.9	0	0	0	0	0	0
3	2.7	18	44	34	0	0	0	0	0
4	0.8	7.5	26	41	24	0	0	0	0
5	0.2	2.8	13	30	36	16	0	0	0
6	0.07	1	6	18	32	30	11	0	0
7	0.02	0.3	2.5	9.7	22	31	24	8	0
8	0.00	0.1	1	5	14	26	29	19	5

Note: A handful of different material filling the first flight is spread and linked into many flights at low concentration after only 8 turns of the screws (in %).

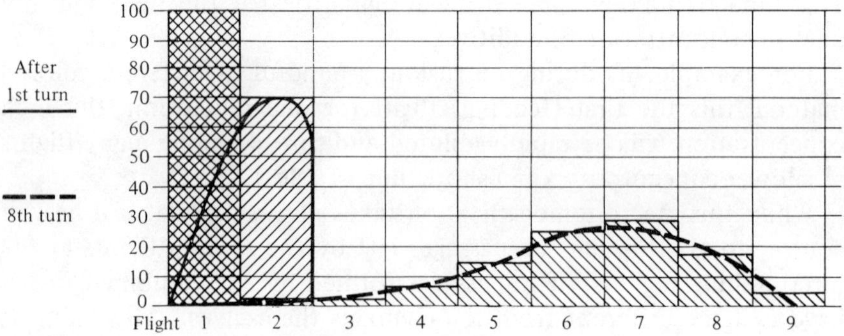

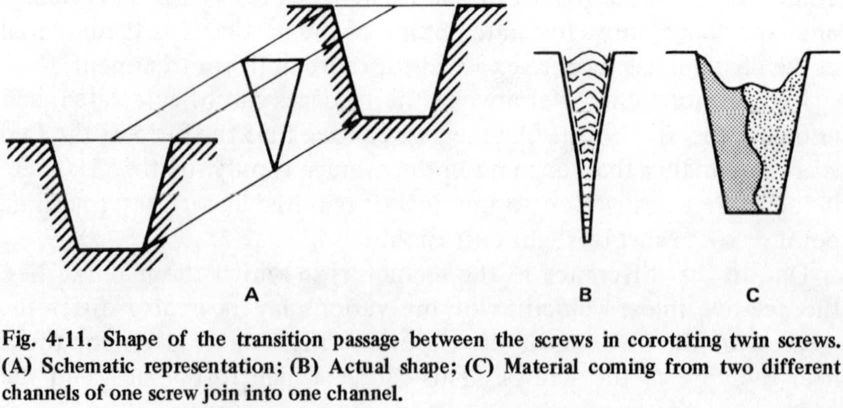

Fig. 4-11. Shape of the transition passage between the screws in corotating twin screws. (A) Schematic representation; (B) Actual shape; (C) Material coming from two different channels of one screw join into one channel.

shear action enhances the disintegration of the particles of material and additives during this kind of internal extrusion (Fig. 4-12).

When, during extrusion, the color of the material fed is suddenly changed, the concentration of the old material in the first flight decreases very rapidly as soon as the feeding is changed over, to become infinitesimal after a few screw revolutions. (See Table 4-2 and 4-3.) This self-cleaning ability of the screws is due to their perfect matching to each other. Conjugation, as we have seen before, means that the flights slide with an angle θ close to the flanks of the channels without touching and with only a minimal mechanical tolerance, practically cleaning each other. Furthermore, in perfectly conjugated screws, the flights in the plane of the screw's axis fit perfectly into

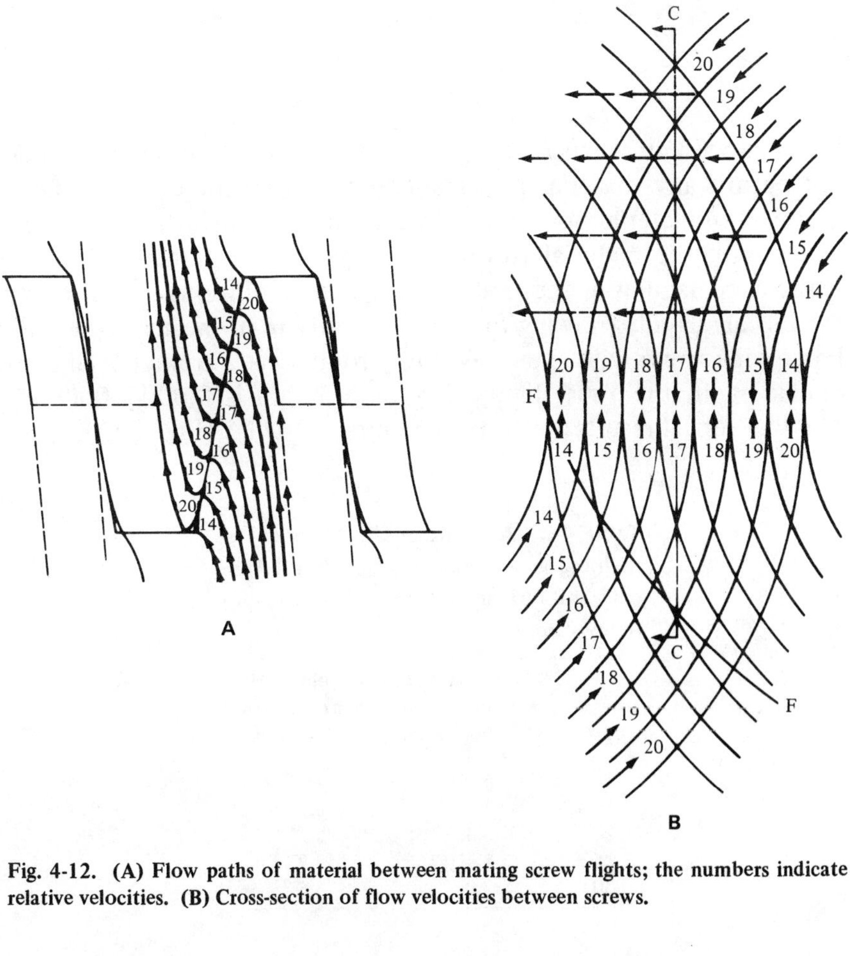

Fig. 4-12. (A) Flow paths of material between mating screw flights; the numbers indicate relative velocities. (B) Cross-section of flow velocities between screws.

Table 4-2. Amount of Old Material Remaining in the First Flight After a Certain Time When the Screws Rotate at 20 rpm.

Time @ 20 rpm	Number of screws revolutions	Amount in %
0	0	100%
3 seconds	1	30
6 seconds	2	9
9 seconds	3	2.7
12 seconds	4	0.81
15 seconds	5	0.24
30 seconds	10	5.9^{-4}
1 minute	20	3.4^{-9}
2 minutes	40	1.2^{-19}
3 minutes	60	4.2^{-30}
4 minutes	80	1.6^{-40}
5 minutes	100	5.1^{-51}

the channel and, when rotating, clean them very thoroughly and do not permit any material to remain stuck within the channel. Only that part of material purposely allowed flows through the wedges and mixes with the material from other flights. The percentage of old material remaining in the first flight after n revolutions of the screws can be calculated, knowing the mixing ability m of the screws, as m^n. For mixing screws with .3 mixing ability rotating at 20 rpm, the amount remaining in the first flight after only 1 min is 0.0000000034%. The theoretical results are confirmed by experience (Fig. 4-13).

Table 4-3. Amount of Old Material Remaining in the First Flight After 1 Minute and After 5 Minutes at Various Screws' rpm.

rpm	% of old material remaining after 1 minute	% of old material remaining after 5 minutes
10	5.9^{-6}	7.1^{-25}
15	1.4^{-8}	6.0^{-38}
20	3.4^{-9}	5.1^{-51}
25	8.4^{-12}	4.9^{-64}
30	2.0^{-14}	3.6^{-77}

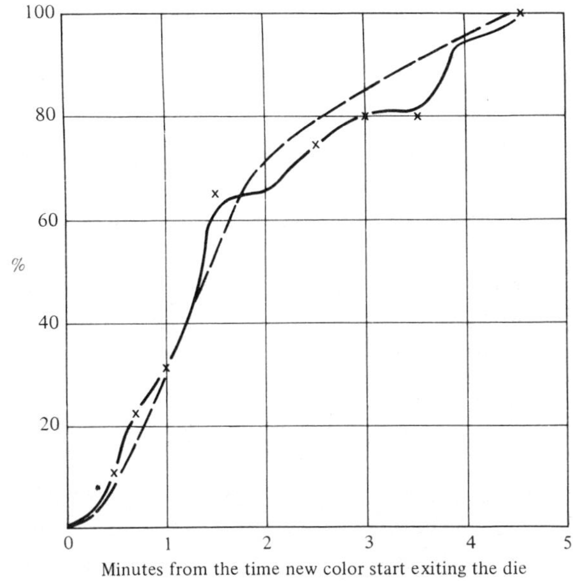

Fig. 4-13. Measured versus theoretical color change (Rigid PVC) in corotating twin-screw extruder.

VENTING

As we have seen, the positive pumping ability of counterrotating twin-screw extruders, makes it easy to establish a point along the barrel where venting can be accomplished. Also, in corotating twin-screw extruders, a section with mainly pumping action is followed by another with very large volumetric capacity where the vent is located and again followed by another pumping section.

With this setup, the venting section has a much lower filling ratio, and the material stays at the bottom of the channels, thus avoiding flooding of the vent port.

The output capacity of a venting section is not affected by back or leakage flows, and so $Q_c = Q_{max}$, and this section has to be made so that $Q_{max} \gg Q$.

Recalling the formula for Q_c, we see that all parameters as D and h have already been fixed, and n is the same, obviously, for the whole screw, so that we will have to increase the pitch angle ϕ. To mix very well the material undergoing venting, for this section we will choose a flight tip $e < E$ so that every part of the material is exposed to degasification.

While the material in the venting section does not fill the channels of the screws, it is, however, not permitted to remain behind, as the self-cleaning action of the screws does occur even within this section. No particle can stick on the screws, where it would degrade and then from time to time appear in the extrudate as black specks deteriorating the quality of the products.

Finally, a counterflighted section placed before the venting section, provides a seal so that we can safely apply vacuum to the venting port without pulling air through the material from the feed opening.

With material under no pressure, with a good mixing and a strong vacuum applied, a very good venting is assured.

PUMPING

If the flight tip e has been chosen equal to or close to E, we have co-rotating intermeshing screws with a pumping configuration. The positive action of the rotating screws pushes the material through the machine and through the die. Every die, however, like any restriction to the flow, presents a certain resistance to the flow of material; therefore, a certain pressure has to be built up by the screws, flight by flight, in order to overcome this resistance and make the material flow through the die.

The flow created by the pressure is related to it as:

$$P = \frac{Q\,\mu}{K_f}$$

where Q is the amount of material actually fed and extruded, and K_f is the conductivity of the die.

As we have seen, the material in this type of extruder is not in a discontinuous state or closed in separated C-shaped volumes, as sometimes has been stated, but is wound around the two screws: so we do have some pressure backflow as well as some leakage flow around the screws. The same pressure P is the one that creates also the backflow, which is:

$$q = \frac{PK_s}{\mu}$$

where K_s is the total conductance of the screws and is proportional to the magnitude of the passages and clearances.

According to the notation of Figure 2-15, the conductance of the small triangular passage in the shape of a tetrahedron between the screws in the intermeshing region is

$$K_\sigma = \frac{(\frac{1}{2}\sqrt{D^2 - I^2})\,[\frac{1}{2}(P - 2E)]^3}{12\,h/2}$$

where $h/2$ is the average length of it. But for each turn of the equivalent screw, there are two of these passages in a series, so K is twice the above.

The conductance of the thin passage between the screw's flight tips and the barrel can be written:

$$K_\rho = \frac{(2\pi D - \sqrt{2Dh})\,\rho^3}{12E}$$

or the length of one turn of the equivalent screw (width) by the clearance to the third power divided by the width of the tip (length of passage).

The sum of the conductances $K_\sigma + K_\rho$ is equal to the conductance of the screw K_s so that the total flow in a turn is

$$q_\sigma + q_\rho = \frac{(K_\sigma + K_\rho)\,\Delta P}{\mu} = \frac{K_s\,\Delta P}{\mu}$$

where ΔP is the pressure differential across that turn and μ the viscosity of the material within that turn.

More than one flight of the screws is necessary to create the required extrusion pressure P at the die. This flow is highest at the first turn nearest to the die, and further back, P decreases and μ increases until the backflow becomes zero (Fig. 4-14).

The flow across N turns is:

$$Nq = K_s \Sigma_1^N \frac{\Delta P}{\mu} = K_s \frac{(P - P_1) + (P_1 - P_2) + \ldots + (P_{N-1} - P_N)}{\bar{\mu}}$$

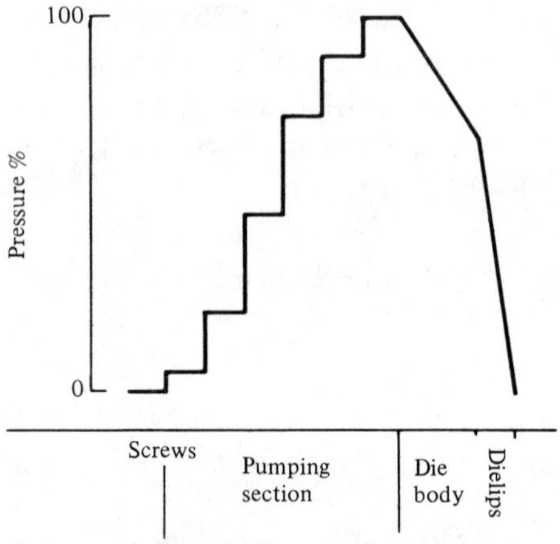

Fig. 4-14. Increment of pressure within the last flights of the screws and the die (measured).

and because the pressure at the Nth flight is equal zero, the flow becomes

$$q = \frac{K_s P}{\bar{\mu} N}$$

where $\bar{\mu}$ is the average viscosity within the filled flights. The pressure P causes the backflow through the first flight in the amount $q = PK_s/\mu$ and through all the filled flights N in the amount $q = PK_s/\bar{\mu}$, but q for continuity must be the same. Equating, we have:

$$\frac{PK_s}{\mu} = \frac{PK_s}{\bar{\mu} N}$$

from which we can see that as far as the backflow is concerned, the number of filled flights N is:

$$N = \frac{\mu}{\bar{\mu}}$$

From the point of view of the screw's design, we have to properly choose those parameters that most affect the total conductivity of the screws in order to reduce it to a minimum. Examining the equation $K_s = K_\sigma + K_\rho$, we note that to achieve this result, E has to be as large as possible, h deep, p and D small, and, most important ρ very small. As screw dimensions have already been chosen for other reasons, it is the clearance ρ between the screw tip and the barrel that has to be minimized.

This clearance ρ between the screws and the barrel, which is critical in single-screw extruders, as it affects the output through increase leakage flow, although it affects the K value in twin-screw extruders, is by far not as critical. In this latter type of extruder the screws are not completely filled with material, and according to the filling ratio, which is determined by the amount of feed, only some of the flights of the pumping section, where the pressure starts to build up, are filled, and few flights at the beginning of this section are only partially filled. If by wear due to long-time use or to work with highly abrasive material the clearance ρ increases, more material will leak backward, and more flights will be filled to reach the necessary pressure; neither the operation of the extruder nor its output will otherwise be affected.

On the other hand, as we shall see, a small ρ means more power used and a high localized shear that we may not want. A larger ρ reduces shear and therefore the heating of the material, increasing its average viscosity, but it also increases the leakage, thereby increasing the number of filled flights. This, in turn, may increase the work put into the material and decrease again the average viscosity, but in this case more gently, as it occurs in a larger number of flights at a smaller rate.

A judicious choice of ρ will balance these results, but most times a not too small clearance is preferable.

NUMBER OF FILLED FLIGHTS

The number of flights needed to create the pressure necessary for extrusion depends, as we have seen, on the resistance of the die.

$$N \equiv Q\mu R_f \quad \text{or} \quad N \equiv \frac{Q\mu}{K_f}$$

where K_f is the conductance of the die.

On the other hand, the clearance around the screws allows for a leakage flow that also affects N.

$$N \equiv \frac{\mu}{\bar{\mu}}$$

so that in final instance we have:

$$N \equiv \frac{Q\mu^2}{K_f \bar{\mu}}$$

or the number of filled flights depends, at a given rpm, on the amount of feed (output), the type of die, and its conductivity and the viscosity of the material. The latter, in turn, is related to the type of material, the heating or cooling of the last part of the barrel, and the shear created by the action of the rotating screws.

For a given die with a certain conductance K_f, an increase in the output, obtained by an increase of the feed, obviously increases the number of filled flights. An increase of rpm, less obviously, decreases it. When the rpm are increased, the shear within the barrel due to the action of the screws increases, and the average viscosity decreases, thereby temporarily increasing N. In the long run, however, as soon as more new sheared material reaches the die, the final viscosity μ also decreases. That the value of this parameter is squared affects N more sharply than the decrease of $\bar{\mu}$, and the net result is a decrease in the number of filled flights N.

This is valid not only for the final pumping section of the screws but also for any pumping section along the screws that also has to build up a pressure to overcome a resistance to the flow, as is the case in a pumping section preceding a counterflighted section.

RESIDENCE TIME

The time required for the material to advance along the screws depends, apart from the screws' speed, on the configuration of the screws themselves, that is, on how many and what kind of sections compose them. Furthermore, it depends also on the degree of filling of these sections or the actual output of the machine.

Each section has different parameters, like channel volume and pitch, that affect the residence time in each section. As long as the material is not melted as in the feed section and the filling degree is way below one, the material will be transported as a continuous flux in the bottom part of the barrel and pushed forward by the screws' flights. When still not melted, the material, especially when the filling ratio is very low, tends to remain on the bottom of the flights, and it advances one pitch per each screw's revolution. The residence time within each of these sections, with length $L = Np$, is, as in the case of counterrotating screws:

$$T_1 = \frac{Np}{V_e} = \frac{N\pi D \tan\phi}{n\pi D \tan\phi} = \frac{N}{n}$$

where N in this case is the total number of flights in these sections.

When the material either fills the screws and is pressed within its channels or starts to melt, it also starts to move around the equivalent screw. Because of the depth of the channels, the mass of it is not affected by the barrel drag, and the rotational speed V_R in this case is $V_R = \pi D n \cos\phi$.

The length of the channel is N times the equivalent spiral, so that we can write:

$$T_2 = \frac{L}{V_e} = \frac{NS_e}{V_R} = \frac{\pi\,(2D - 0.9\sqrt{Dh})\,N}{\cos\phi\;\pi\;D\;n\;\cos\phi} = \frac{2D - 0.9\sqrt{Dh}}{D\cos^2\phi}\;\frac{N}{n}$$

which usually represents approximately:

$$T_2 = 1.63\,\frac{N}{n}$$

The second equation is valid not only for the pumping sections but also for every section after the material has melted. This can be verified on a model with a transparent barrel or just by looking inside a vent hole where the material is already melted but by far not filling the channels, lies against the bottom of the channels and the front flank of the flights, as is shown in Figure 4-15, and can be seen rotating with the screws. Practical tests have shown the value of residence time, per turn, in sections in which the material is not yet melted, to be approximately equal to $1/n$ and in the sections where the material is melted, to be approximately $1.63/n$; this is equivalent, respectively, to 3–5 and 5–8 sec per turn in the average praxis.

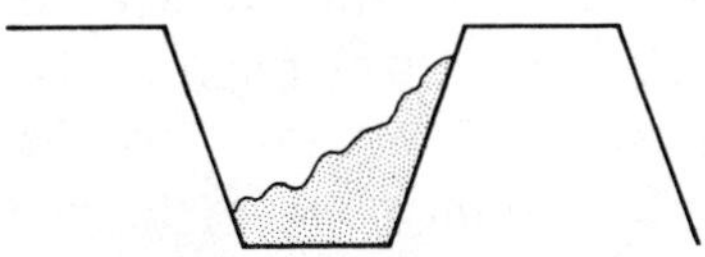

Fig. 4-15. Melted material in a venting section with very low filling ratio.

The residence time before the point where the material melts is of importance in the design of the extruder, as during this time the material should be preheated in preparation for melting. After that point, the residence time is important only because a too long residence time at high temperature may affect thermally unstable resins. However, what is important is not the average residence time of the material within the screws but actually the longest time some portion of the material will remain in the barrel. In a small single-screw extruder ($2\frac{1}{2}''$ at 60 rpm), the average residence time may be between 1 and 3 min but some material may issue from the barrel after 30 sec and some after 10–15 min. In corotating twin-screw extruders – because of their positive pumping action and their self-cleaning ability – the average time is practically the same for all the particles of material, and none of them will stay behind for a longer time with the risk of decomposition or degradation.

The residence time of the melted material will be as long as needed for the material to go through the various sections, each performing a different task. In a modern extruder with screws composed of many sections each with 3–6 turns and some with 8–10 turns, the residence time before melting can be approximately 70–100 sec and within the area where the material is melted another 90–150 sec, which brings the total residence time to 160–250 sec, or, depending on the screws' configuration, between 2.6 and 4.1 min. Then the flow through the barrel head, adaptor plate, and die, according to the die size and output, will add another 10–12 min, for a total of approximately 15 min before the material leaves the extruder.

PRESSURE PROFILE

The early twin-screw extruders that were built had screws simply composed of feeding, mixing, and pumping sections. In those extruders, the filling ratio was way below one and pressure could have been, and was, built up only in the last few flights of the pumping section. With the introduction of the reverse-flight section, flights could be filled and pressure built up according to need in any place along the barrel. As we have seen, the direct flight section, which pushes the material forward against the flow caused by the counterflighted section, has to build up a certain pressure to make the material flow through the slots of the counterflighted section.

This pressure is built up flight by flight, so some flights are required to reach the necessary pressure. The counterflighted section following a pumping section is completely filled with material, and as the material proceeds along the slots, the pressure in them drops. We have then in the part of the screws encompassing a pumping and a counterflighted section a pressure profile according to Figure 4-16.

The amount of pressure that can be built up depends on the ratio of Q_c of the pumping section to the Q_c of the reverse flight and from the width of the slots cut in the counterflighted section. Peak pressures of 200–270 atm (approximately 3000–4000 psi) have been reached. The pressure at the end of the counterflighted section depends on the length of it and on the configuration of the next section.

An increase in rpm does not change the situation, as it increases the output capacity of both sections by the same amount, but, because of that, it reduces the number of flights needed to reach the required pressure. On the other hand, an increase in output, given by an increase in actual feed, increases the pressure needed for the material to flow through the slots as $P \equiv Q/A$ but also increases the number of flights filled, as N is also proportional to Q. For the same couple of pumping and counterflighted sections, we have then the possibility of having the pressure build up according to any of the curves shown in Figure 4-17.

If it is required by the type of extrusion performed, a second set of counterflighted sections can be placed farther down the barrel.

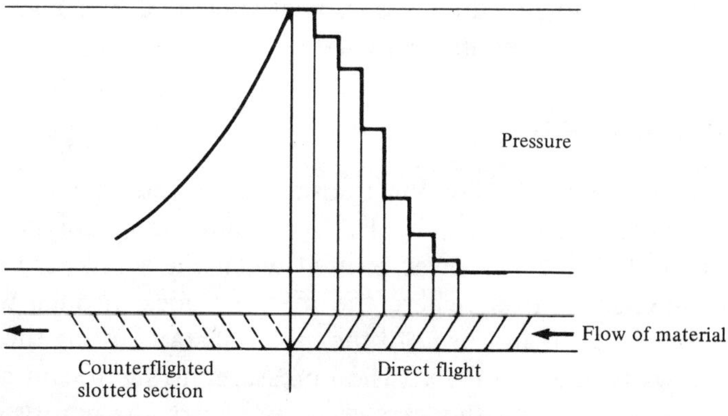

Fig. 4-16. Pressure buildup in a direct flight due to the presence of a counterflighted section.

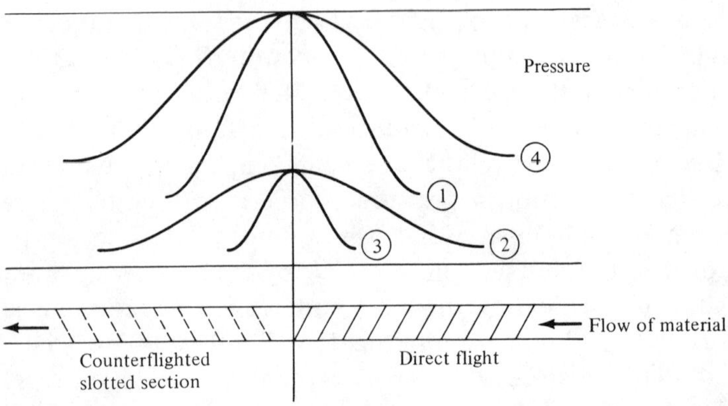

Fig. 4-17. Pressure buildup: (1) high Q, high rpm; (2) low Q, low rpm; (3) low Q, high rpm; (4) high Q, low rpm.

This can be necessary in the case of adding a gas to the melt mid-barrel, as in the production of expanded Polystyrene or Polyethylene foam. The two counterflighted sections create two high-pressure sealing zones and between them a chamber where the pressure is low. There easily can be injected the gas needed for foaming.

A second counterflighted section may also be necessary when, after a zone where the material is under no pressure, we want to re-compress the material or homogenize it well.

According to need, the pressure within the barrel can have almost any desired profile. This can be established beforehand by chosing the proper configuration of the screws.

A possible example of the pressure profile created by a given screw configuration can be seen in Figure 4-18.

POWER CONSUMPTION

Except for the amount needed to transmit the torque from the motor to the screws through the mechanical gear train and rotate the screws, a very small amount of power is used when the screws are empty. With an open-end barrel, where the material goes through without any pressure buildup and/or until the material starts filling the flights of the screws to create the extrusion pressure, no significant amount of power is used above the one needed to turn the empty screws. Power consumption starts to be significant only when some flights are filled up.

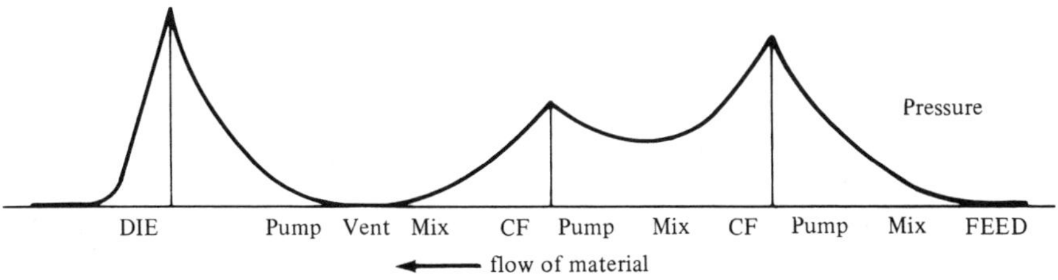

Fig. 4-18. Example of pressure profile in the barrel with a screw configuration having two counterflighted sections (CF) and a venting section.

When channels are filled with material, shear is developed, and a corresponding amount of power is thereby dissipated. The amount of power used for shear in the channels is

$$Z = \mu \dot{\gamma}^2 \; V = \frac{\bar{\mu} \pi^4 D^3 De \tan \phi}{2h} n^2 .$$

This is the same as in counterrotating screws and is valid for the nonintermeshing parts of the flights.

In corotating screws, however, not all flights are filled, and the result of the formula has to be multiplied only by the number N of filled flights and not the total number of flights in this section.

Low rpm, as used in all twin-screw extruders and deep flights, keep $\dot{\gamma}$, and therefore the power dissipated as shear within the channels very low.

Examining again the screw's geometry we can see that by far the largest part of the remaining power is used by the milling action that occurs (1) between the flight tip and the barrel, (2) between the flight tip of one screw and the screw root, or bottom of the channels of the other screw, and (3) between the flanks of the flights in the intermeshing region.

These actions can be expressed with the following equations. As far as (1) is concerned, the flight tip, of a width e and length equal to one turn of the circumference of the equivalent screw, at a distance ρ from the inside surface of the barrel, moves with a speed v in relation to the barrel. As we have seen before, the first equation can be written:

$$Z_c = \frac{\mu v^2 e C_e}{\rho} .$$

This has to be multiplied by the number of turns N where the material fills the screws, or, which is the same, by the number of turns that creates the extrusion pressure.

Because $v = \pi \cdot D \cdot n$, we can now write:

$$Z_c = \frac{\pi^2 D^2 e C_e}{\rho} \mu n^2$$

which is the same formula as in counterrotating screws.

The second point of shear is in the slot of thickness ϵ left by the mechanical clearance between screw tips and channel bottoms, which, in corotating screws, move in opposite directions. The speed of the screw tips is $\pi \cdot D \cdot n$ and that of the screw root is $\pi Dn - 2h$; the combined speed is therefore:

$$v = 2\pi n\,(D - h).$$

Starting from the roller mill formula, we can disregard, as we did before, the thickness of material behind the slot, as it is much larger than the clearance; also, taking into account that the tips are on a diameter D and the roots on a diameter D_i, we can take as a roll diameter the mean value $D - h$ which is the interaxis I. Again, per each turn, the screw profile involves one tip and one bottom of the flights; therefore, the length of the slots is $2e$. We can then write:

$$Z_s = \frac{\mu v^2\,(D - h)\,2e}{\epsilon}$$

or, substituting:

$$Z_s = \frac{8\pi^2 I^3 e}{\epsilon} \mu n^2$$

This is the point of major shear in corotating screw extruders. To limit the shear and without basically changing the motion of the material within the screws, a gap twice as large as anywhere else is usually left here by slightly reducing D_i.

A third point at which shear is developed is at the flanks of the screw flights; there again the basic formula is the same:

$$Z = \frac{\mu v^2 A}{\sigma}$$

but here the speed is the speed of the two facing surfaces, which has to be added to one another as they move in opposite directions. The average speed at the median diameter $I/2$ is: $\pi\frac{I}{2}n$, so $v = \pi In$.

The area involved here, as in counterrotating screws, is

$$\frac{1}{2}h\sqrt{D^2 - I^2}$$

and the distance between the flanks is σ. So we can write:

$$Z_\omega = \frac{\mu\pi^2 I^2 n^2 h\sqrt{D^2 - I^2}}{2\sigma}$$

or

$$Z_\omega = \frac{\mu\pi^2 I^2 h\sqrt{D^2 - I^2}}{2\sigma} n^2$$

Remembering what was previously said about the area of interference and about the choice of e, we can see that when $e = E$, we have one side of the flanks sliding along the flanks of the other screw. When $e < E$, the very wide passage resulting cannot be treated in the same way.

Finally, beside the power dissipated into the material as shear, we require power also to create the pressure P needed for extrusion. This is:

$$Z_p = P \cdot Q.$$

Substituting the conductance of the die, we have:

$$Z_p = \frac{Q^2\mu}{K_f}.$$

The total power transmitted from the main motor to the screws is therefore:

$$Z_t = N(Z + Z_c + Z_s + Z_w) + Z_p, \text{ or}$$

$$Z_t = \left[\frac{\pi^4 D_e{}^3 D \tan\phi}{2h} + \frac{\pi^2 D^2{}_e C_e}{\rho} + \frac{8\pi^2 I^3 e}{\epsilon} + \frac{4\pi I^2}{\sigma}\right] \bar{\mu} N n^2 + \frac{Q^2\mu}{K_f}.$$

Once the screws are designed and established, as the output is separately controlled, the power required is a function of the following parameters:

$$Z_t = f\left(\frac{Q^2 \mu}{K_f}; \bar{\mu}; N; n^2\right)$$

From the point of view of operation, we can also see that among the factors influencing the power consumption during extrusion are the average viscosity and the number of flights filled. However, as N is inversely proportional to the average viscosity within the filled flights $\bar{\mu}$, these two factors somewhat compensate each other. So the most important factors are the screw speed and the actual output, which affect the power consumption by the square of their value.

Comparing this with the equation Z_t for counterrotating screw extruders, we note that the difference lies in the power dissipated as shear at the flanks of the flights and that here σ affects the power linearly and not, as in counterrotating screw extruders, by the cube of its value.

Applying the above formulas, we can calculate the power used in corotating twin-screw extruders.

We take as an example the extruder type RC 9, manufactured by LMP Colombo of Torino, Italy, which has the following data:

$$I = 6.6 \text{ cm}; D = 8.2 \text{ cm}; h = 1.6 \text{ cm}; p = 3.14 \text{ cm}; \phi = 6°57'; e = 0.9 \text{ cm}$$

and has a constant torque motor with 3.4–15.65 hp variable with the speed, rotating the screws between 9 and 34 rpm. The clearances σ and ρ have been taken as 0.06 cm and $\epsilon = 0.12$ cm. Operating this machine at 20 rpm or 0.333 sec^{-1}, the value of Q_c results 21.6 cm^3/sec, for a real maximum output of $Q_c - q$ = 18.8 cm^3/sec, corresponding to 148 lb/hr.

At this output and with a regular 1 x 1 cm die, we can see that approximately 10 flights are full, and within these flights the calculated power consumption is as shown on the top of p. 91.

Z	Shear within the channels	0.3454	
Z_c	Shear between tips and barrel	0.9955	
Z_s	Shear between tips and bottoms	4.0700	
Z_w	Shear between flanks	0.6666	
Z_p	For pressure buildup	0.2530	
Z_t	Total	6.3305	75%
	Power for drive	2.11	25%
	hp	8.44	

At 20 rpm, the main motor is rated 9.2 hp, which leaves 0.76 hp or 8% leeway for peak absorption.

In this extruder the motor, however, contrary to what occurs in single-screw extruders, does not furnish all the energy required to melt the material. We know from page 51 that to bring the material from room temperature to extrusion temperature, we have to increase its enthalpy by approximately 0.1 hp/lb, according to temperature and material (Bernhardt mentions 0.08 to 0.15 hp/lb/hr), which would require for 148 lb of material 14.8 hp. In this case, the motor supplies only 6.33 hp to the material, or 42.8% of the needed energy; the rest, or 8.47 hp, must and is supplied by the heaters around the barrel. This corresponds to 6.31 kW or very approximately one-half the 15 kW of the power of the heater bands installed.

5
Construction Features

MOTORS

Twin-screw machines are basically built very similarly to single-screw extruders; the components, at least, are the same. Even here we have a body or basic frame containing or supporting and connecting the motor, the gear box, the feeding system, the barrel, and the screws.

In all twin-screw extruders, whether corotating or counterrotating, the main motor supplies only approximately one-half of the required power; therefore, being comparatively small, the motor can be placed under the barrel or within the basement of the extruder. This can save valuable space without cluttering the area around the machine, enhancing safety, cleanliness, and simplicity of operation.

Some of the very old machines had AC motors with fixed synchronous speed and an automobile-type gear box to shift gear, which resulted in three or four prefixed speeds for the screws.

Subsequently, the use of belt-type speed variators (PIV) enabled the screws to have an infinite rpm setting, while the motor remained a fixed synchronous AC motor.

Today, variable-speed DC and AC motors, and in rare cases hydraulic motors, are used.

Better and more efficient electronic controls have widened the use of DC motors, especially in the United States, whereas in Europe commutator AC motors (like the ones manufactured by ASEA, Sweden, and by their subsidiaries around the world) are preferred.

This last type, known also as the Schrage-type motor, is very often the most economical from the point of view of both initial investment and operational cost. Basically, these are three-phase AC motors in which the stator is fed through a commutator; depending on the position of the brushes, the rotor can be caused to run at subsynchronous or supersynchronous speed with a range of n_{max}/n_{min} from 3 to 8 or sometimes up to 17 or 20.

SCREWS

We have seen that in twin-screw extruders each section of the screws performs primarily a specific operation. According to the design of the screws, some manufacturers make them in one piece, whereas others make them in segments. These segments can be interchangeable so that each one can take the place of any other; in this case, the shaft is continuous with a spline, or key, on which these segments can be assembled in any desired order. Or they can be exchangeable so that each segment, usually a functionally complete section, can be exchanged for a new one of different characteristics but cannot be placed anywhere else in the screws; in this case, each section has its own splined tail that joins the preceding section. According to needs, the screws can sometimes be cooled. So, in many instances, the screws are cored and equipped with oil circulation. If the oil is brought through an inner pipe to the end of the screws and returns around it within the core of each screw, then the cooling will be more effective at the screw ends or tips and less at the rear sections. If the inner pipe contains the return flow, then the reverse is true. For screws with a continuous shaft, this type of cooling is efficient only at the screw tips; for screws made in sections, its effectiveness in each of them depends on the wall thickness of the section and the volume of the chambers where the oil runs. Even in the case in which the oil return occurs through an inner pipe, generally speaking, a better cooling occurs at the screw tips.

Due mainly to the sometimes very low bulk density of the material fed, it is in the first (rear) part of the screws where the channel volume has to be maximized. As we do not want too large a pitch, for such a pitch would decrease residence and therefore heating time, it is the channel depth that has to be increased. Because the screws are intermeshing and conjugated, and their interaxis is fixed, the larger the channel depth h, the larger the screws external diameter D. This results in screws with larger diameter in the rear than in the front; for cleaning, the screws cannot be extracted through the front of the barrel, but it is the barrel that is to be pulled away toward the front (tips) of the screws.

GEAR BOX

In all twin-screw extruders, obviously, the gear box is more complex than in single-screw extruders, as the motor's power has to be transmitted to two shafts. Corotating twin screws require one more inner gear to reverse rotation. However, in modern extruders, gears and bearings – sometimes computer calculated – well lubricated by forced circulation of suitable oil, do not present any problem, provided, of course, that the torque limits are observed.

As far as mechanical construction is concerned, twin-screw extruders do not require that the elements of the drive be as massive as in corresponding single-screw extruders because, as we know, much less power and smaller motors are involved. In many cases, what seems a thin, fragile shaft to an eye used to single-screw extruders is in reality a solid, even oversized piece of machinery. Furthermore, twin-screw extruders do not require the same thrust-bearing capacity as single-screw machines because of their different operating principle. Contrary to the latter, twin-screw extruders develop just the correct pressure required to force the material through the die.

THRUST BEARINGS

One of the major problems inborn in twin-screw extruders manufacturing was how to support the axial thrust on the screws because their limited distance between the screws shafts. Due to the limited interaxis, large diameter conventional thrust bearings cannot be used and, since the capacity of a thrust bearing is proportional to its diameter, nor could small bearings. Other systems had to be found.

One solution since early times was to stagger the two thrust bearings so as to use the maximum possible diameter bearings (Fig. 5-1).

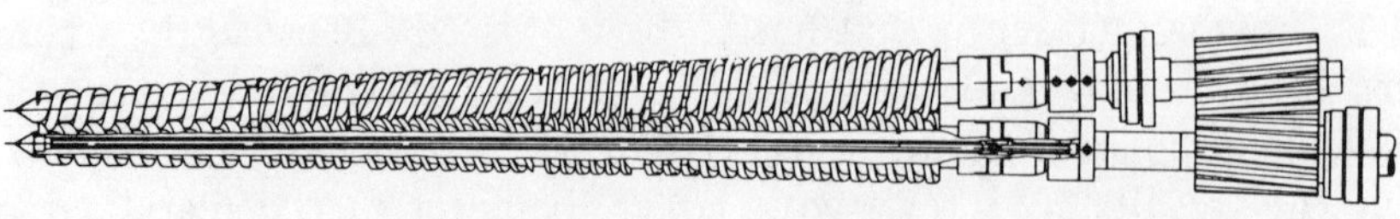

Fig. 5-1. Thrust bearing arrangement (Cincinnati-Milacron).

Then the use of multicollar thrust bearings was introduced; this consist in a series of rings on the shafts engaging into circular grooves in the support, oil lubricated, each of which supports part of the load. This type of bearing, however, requires a perfect machining and alignment of the various pieces (Fig. 5-2).

An intermediate solution is to use one of these multilayer bearings for one shaft, and at the end of the other shaft, made a little longer, placing a normal large size thrust bearing (Fig. 5-3).

Another solution to this problem is the oleostatic pressure bearing (CAFL). With it, the end of the screw shafts rotate with close tolerances in sleeves so as to create chambers in which a fluid is injected whose pressure equals the back pressure of the screws. When this back pressure increases, the slight axial movement of the screws backward due to the imbalance between back pressure and oil pressure controls the size of the oil-incoming flux opening, thereby increasing the oil pressure and reestablishing the balance.

Although all these systems work well, the innovation that really eliminated all thrust-bearing problems was the multilayer oleodynamic system.

This type of bearing is derived from the Mitchell-type pressure lubricated axial support. The Mitchell bearing consists simply of two parallel plates, one of which has some sort of movable "shoes"

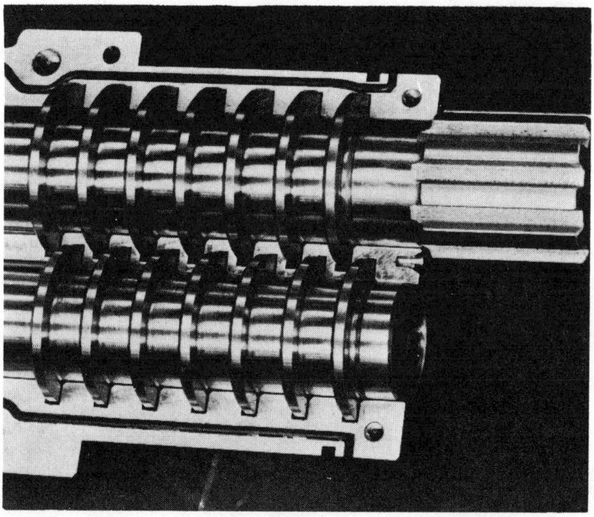

Fig. 5-2. Thrust bearing with collars and grooves (Schloemann).

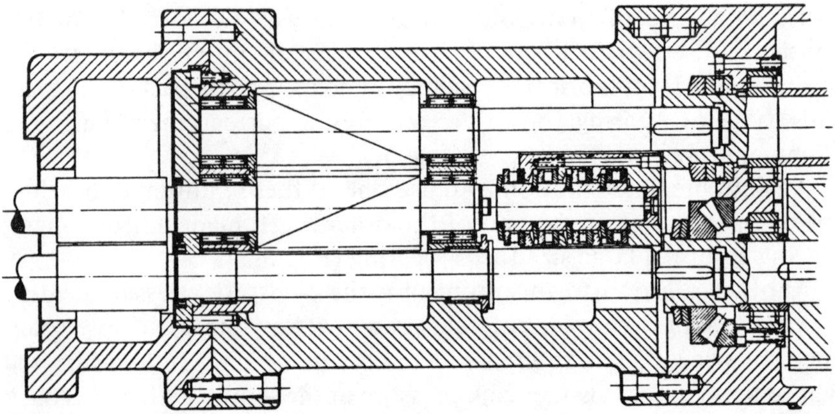

Fig. 5-3. Thrust bearing mixed arrangement with one large and one multiple small bearing (Anger).

that float on a veil of oil maintained in between and orient themselves so as to create forces that keep the plates apart. Their functioning does not rely on the oil pressure, but their action is dynamic or relies on the motion.

In their simplified version adopted on extruders, they work as follows: as a substitute for the floating "shoes" in one of the plates of these bearings, there are radial grooves that are deeper on one side and smoothly reach the surface on the other (Fig. 5-4). During rotation, these slopes push the oil against the other plate, which is flat, and ride on the veil of oil so that one plate never touches the other.

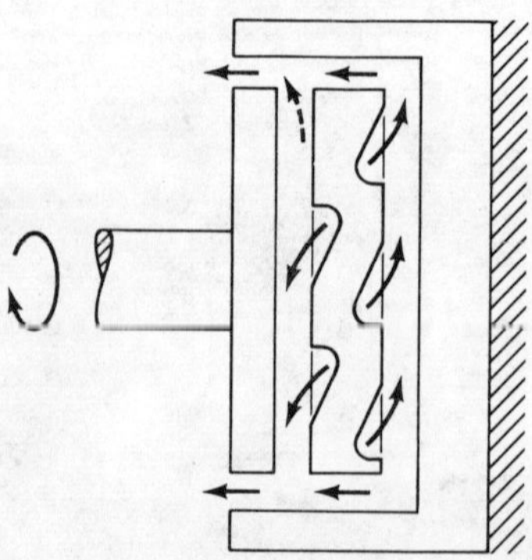

Fig. 5-4. Principle of operation of the oleodynamic thrust bearings.

Practically, they are composed of one set of stationary rings, one set of rings splined to the rotating shafts, and a set of free discs placed between them. These rings have the radial grooves that create the oleodynamic action. (Fig. 5-5)

As long as the oil, injected through holes in the shafts, is present, the oil bears the load and transmits it to the fixed plate (Fig. 5-6). A loss of oil, however, means instant contact between the rotating surfaces and an immediate ruining of the bearings. For this reason, the electric circuitry must provide a safety circuit for immediate stoppage of the machine if lack of oil pressure should occur; moreover, the oil-circulating pump usually starts first, and the main motor driving the screws will only start after the correct oil pressure is achieved.

One obvious advantage of these bearings is that, insofar as there is no contact between the facing surfaces, there is no wear; therefore, they can last practically indefinitely.

Another advantage is that they can easily be placed in a series without difficulty, as variations in the thickness of the veil of oil take care of possible minor differences in the thickness of the metal components, thus perfectly equalizing the load among them.

Fig. 5-5. Oleodynamic thrust bearings consisting of two stationary (large), two rotating rings (small), and two grooved discs per shaft.

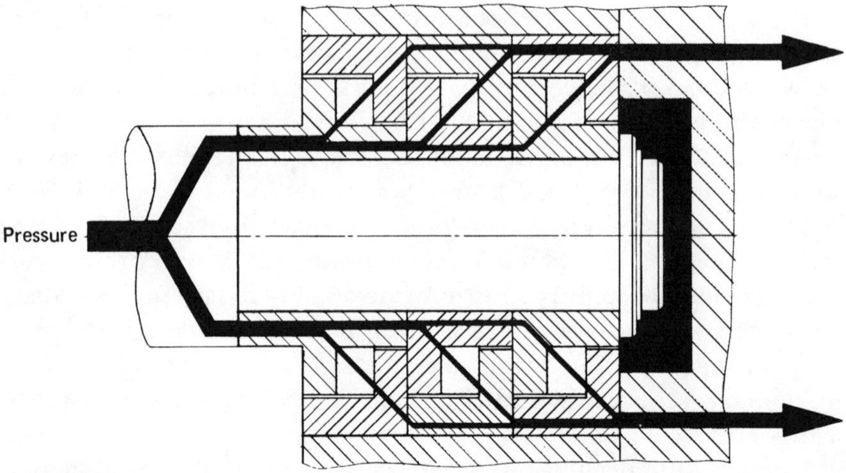

Fig. 5-6. Oleodynamic thrust bearing (LMP/Colombo) showing transmission of axial load from shaft to fixed plates.

These bearings, used in extruders for more than 30 years, (LMP/Colombo, Windsor) have proved themselves in continuous operation for as long as 15 years under pressures of 3,000–4,000 psi.

FEEDING SYSTEMS

Another difference between many twin-screw and standard single-screw extruders is the use, by many of the former, of a feeding system.

This is a separate mechanism driven by a variable-speed motor that measures the amount of material fed and literally drops it into the barrel.

These mechanisms can be of various types, either volumetric, which measures by volume, or gravimetric, which measures by weight; whereas this last type is more complex and costly, it is preferred when it is necessary to have a precise dosage of the polymer and of the additives. In this case, two gravimetric feeders, one for the polymer and one for the additives, give the mixture the right proportions. Gravimetric feeders are also used when making foam by the direct injection method, as a precise ratio of polymer and gaseous blowing agent is necessary.

The simplest of the feeding mechanisms is the auger or conveying screw. This consists of a hopper, which contains the material to be fed into the extruder, at the bottom of which is a deeply flighted screw of high volumetric capacity. This auger, rotating very slowly, transports the material into the extruder feed opening and controls its flow at a rate that is proportional to the auger's speed.

Although the flow of material fed into the extruder is somewhat pulsating, the overall amount is pretty much the same for any given period of time; the extruder's screws take care of distributing it and average the flow among their channels, so that a few screw flights past the feed opening, the flow of material is regular and constant. Even if the feed is not continuous but somewhat pulsating, as long as it is constant in time, the system is satisfactory.

Among the gravimetric feeders, the more complex and more precise move an uninterrupted stream of material on a conveyer belt and weigh it while it moves so that a steady flow of x gm per each second are fed into the extruder.

Other types of gravimetric feeder mechanisms fill a bucket mounted on a scale, and after this is filled to the correct weight, drop the material into the extruder; this sequence takes a few seconds. Once weight and time have been set, these systems drop, for example, 376 grams every 20 seconds; these successive bunches of material are equalized by the first section of the screws, which transform them into a continuous stream of 18.8 gm/sec or 148 lbs/hr. All these types of feeders are commonly used for other purposes as well and are available on the market.

The advantage of having a separate means of controlling the input of material into the extrusion screws cannot be overemphasized; it separates screw speed, and therefore shear, from the output of the extruder and gives full control on the work that the material undergoes.

BARREL

The length of the barrel in twin-screw extruders has, as we have seen, nothing really to do with the way these machines work, but the barrel is only as long as it is required by the configuration of the screws. In these machines, the barrel is usually much shorter than in single-screw extruders, and the L/D ratio has no real meaning with them, at least in the usual sense. However, a certain amount of length of barrel is required to transfer the necessary heat from the outside of the barrel into the material, and this, as we have seen, is very important in twin-screw extruders.

The heat, provided by heater bands, cannot exceed a certain number of watts per unit of surface in order to limit the temperature of the heating wires and prolong the life of the heaters. If this specific heat density remains the same, extruders with larger output requiring larger amount of heat must have the part of the barrel before the point where the material must be melted much longer than smaller machines. As the output increases with the square of the screw's diameter *D*, so is the surface of the barrel expressed in *D* lengths; therefore, the same number of diameters is required to furnish the required heat. For example, for a 3″ *D* screw, if a length of 15″ or 5 times 3″ is required, for a 4″ *D* screw, a 5 times 4″ or 20″ of barrel length is required.

The other parts of the barrel are little affected by the output, and only very little lengthening is required in the other zones. Because of this, large capacity extruders with large screw diameters tend to have actually only slightly larger L/D ratios than smaller machines, varying from 12–15 for straightforward screws up to 19–22 L/D for more sophisticated screw configurations.

The tremendous barrel length of some single-screw extruders, which can reach 30 or even 50 *D*, in twin-screw extruders is not only unnecessary but would actually be counterproductive.

Beside the above, corotating twin-screw extruders differ also in the screw and barrel construction. As in these extruders, the material completely surrounds the screws and is under the same pressure all around their circumference. The screws are centered within the barrel by the material itself. This avoids the rubbing of the screws against the barrel since, during normal operation, there is no contact between screws and barrel. In contrast, in counterrotating twin screws, there is a pressure differential around the screws that pushes them apart and against the barrel's inner wall. Therefore, in corotating machines, there is no need to line the barrel with special alloys or to reinforce the tip of the flights with wear-resistant material.

ELECTRICALS

Basically, all electrical circuits necessary to run an extruder are motor circuits, heater circuits, and auxiliary circuits.

Motor circuits are simplest when AC motors are used, as these can be directly connected to the main lines. DC motors require a system to change the alternating into direct current. This not only renders the circuitry more complex but also may be, as it is with any added component, an extra source of trouble.

Apart from the start-stop buttons and speed-regulation control, an ampmeter is always included in the motor circuit; although only a wattmeter would indicate the real power taken by the motor, the ampmeter permits an accurate enough watch on the machine work. An rpm indicator is also an integral part of the motor's control.

Electrical heating is most important in twin-screw extruders, for approximately half of the energy needed is furnished by it. Heater bands, fitted onto the barrel and divided into zones, usually corresponding to the various screw sections, heat the barrel, each one furnishing the heat required in that zone along the barrel.

Although modern heater bands, especially the cast-in type, have a long life and can have a high specific load, usually, for reasons of durability, the load is kept between 2.5 and 3.5 W/cm^2, with some exception up to 4 W/cm^2.

Another type of electrical heating for the barrel is the induction system. AC current circulating in coils around the barrel set up a magnetic flux within the barrel, and the Eddy currents generated by it directly heat the barrel itself. High cost aside, this type of heating offers many advantages in durability, ruggedness, and low power consumption.

Electrical heating is measured by thermocouples or thermistors fitted within the barrel's wall that, in turn, operate temperature controllers.

Whereas in single-screw extruders barrel temperature must be controlled within very narrow limits in order to keep the layer of material nearest to the barrel's inside surface within the adhesive range (otherwise, slipping and pulsating may occur), in twin-screw extruders, there is not such a strict requirement. A simpler controller can be used, and even if the barrel temperature should wander plus or minus a few degrees from the one set, this will have no influence on the extrusion. As in twin-screw extruders, it is not a must that the material's temperature remain within the narrow limits of the adhesive range.

Proportional controllers, where the power supplied decreases when the indicated temperature nears the set temperature, are certainly of advantage, although simple on-off controllers can be used in many cases.

Except when the extrusion temperature should be lower than melting and mixing temperatures, in twin-screw extruders, there is no need for cooling devices. If there is such a need, a separate system, usually by oil circulation, provides cooling to the screws and the barrel.

Other auxiliary motors, as for feed, for forced lubrication, etc., have simple and straightforward circuits.

One important device that some twin-screw extruders have is a safety interconnecting circuit. This is to avoid pitfalls like starting the feed before the extrusion screws are running, or the reverse, stopping the extrusion screws, leaving the feed on. If this, by mistake, should occur, an excessive amount of material would fill the first section of the screws, and when these are restarted, a plug of material would be propelled through the machine without the machine having a chance to properly melt it. When it reaches the die, trouble may occur.

In machines with oleodynamic thrust bearings, this safety interconnection is essential, for when the start button is pressed, it is not the main motor that is started but the oil circulation pump. When the system is assured that lubricating oil is present and under pressure, then, and only then, the main motor will start, thus assuring the safety of the thrust bearings.

Being very compact and simple machines, twin-screw extruders usually come already prewired, and only simple connections between the main lines, the control cabinet (if any), and the extruder itself have to be made.

TORQUE

The maximum power required at the screws can be calculated by inserting in the power formula given above (see pp. 59 and 89) the maximum value of $Q = Q_c - q$ and the appropriate values of viscosity and K_f. The torque applied to the screws' shafts will be:

$$M = \frac{Z_t}{n} \quad \text{or} \quad M = 716.2\,\frac{\text{hp}}{\text{rpm}}$$

As we have seen, Z_t varies very little with changes in rpm; the torque instead varies in inverse proportion to the rpm. We have seen previously (p. 50) that, depending on their specific heat, the total specific energy needed to melt the material can reach 0.167 kWh/kg, (at .8 Cal/kg and a $\Delta t = 180°C$) and as only approximately one-half of this energy has to be furnished by the screws, these should be able to supply one-half of it, or 0.0835 kWh/kg. The motor, through the screws, must be able to transmit to the material approximately 0.1 hp/lb/hr produced.

If we take this maximum value and multiply it by, let's say, a safety factor of 3, we can then calculate the maximum M by inserting the lowest n into the above equation. In twin-screw extruders with independent feeding, this can be as low as 12 or even 10 rpm.

Once the maximum specific torque has been calculated, knowing that each shaft transmits only one-half of the total torque, we can determine the diameter of the screw shafts as:

$$d = \sqrt[3]{\frac{16M/2}{\pi \;\tau}}$$

where τ is the resistance of the steel to tangential stresses. The diameter of the main shaft driving both screws will be $d_1 = d\sqrt[3]{2}$, or 1.26 times the diameter of each shaft.

In the example calculated on page 91 the total Z_t is 6.33 hp. The Z_t on each shaft is then one-half of it, and by taking into account a safety factor of 3, the torque on each shaft is then:

$$\frac{M}{2} = \frac{1}{2}\, 716.2 \;\frac{3 \times 6.33}{20} = 340 \text{ kgm.}$$

Assuming for a good-quality alloy steel a τ of 22 kg/mm^2, the diameter of each shaft is $d = 4.28$ cm. Since it is in the feed section that the screw shaft has to withstand the maximum torque, it is there that it should have at least the calculated diameter. This diameter corresponds to the diameter D_i of the screw's root.

In the machine examined in the previous example, (see page 90) the screw's root D_i is, in fact, 5.0 cm.

These calculations can therefore verify whether screw size, power, and torque requirements correspond to each other.

Calculation of power and torque used with a certain material and screw configuration can also give us an idea of what production output we can expect from that machine, as the total admissible torque is usually given in the machine specifications.

6
Operating of Twin-Screw Extruder

Working with twin-screw extruders is much different from operating a single-screw extruder because of the widely different concepts on which they are based.

The operator therefore has to be trained to handle these extruders properly, and sometimes the experience acquired with single-screw machines is more of a hindrance than an advantage.

First of all, it must be kept in mind that the positive pumping action of the screws in twin-screw machines will propel the material to the die and try to force it through its opening no matter how fluid or highly viscous it is. Consequently, care has to be taken that the material is properly plastic at the moment of extrusion. As any extruder, twin-screw machines have to be preheated before starting. Because of the polytropic nature of these machines, a large part of the heat for melting is given not from the shearing action of the screws but through the barrel; therefore, good preheating is most important. Not only do the right temperatures have to be reached, but also some time has to be given to allow the heat to sink from the outside of the barrel to the screws and bring them also to the right temperature. As the temperature is measured usually at the barrel's outer part, it will show there a higher value than the screw's temperature. It is therefore useful, at least for a certain period before start up, to set all temperatures along the barrel slightly higher than requested and lower them after the machine has been started.

Furthermore, it makes a great deal of difference if the material is directly fed into the screws or if it is metered into them by a separate device. In the first case, the output is proportional to the screw's speed; shear, and power consumption are proportional to the square of the screw's rpm, as occurs in single-screw extruders; when a separate mechanism controls the amount fed, the screw's rpm has nothing to do (within limits) with the amount of material extruded and only controls the shear and the power dissipated into the material by the screws. With these points in mind, a twin-screw extruder is easy to operate.

START-UP

From the operational point of view, we must differentiate between self-feeding and metered screws.

When the material is directly fed into the screws, for the above-stated reasons, the extruder can be handled basically, at start-up, as a single-screw extruder. When, however, the feeding of the material is separately controlled, as in corotating and some counterrotating twin-screw machines, the rules of operation are quite different; it must be remembered that after start-up the material can be fed only at a low rate in relation to the screw's rpm so that more heat will be available in the barrel per unit of weight of the polymer. To better heat the material during start-up, the residence time, which, as we have seen, is inversely proportional to n, can be kept high by starting at somewhat lower rpm than usual.

This will also increase the filling ratio for the amount of material fed. As for a given die:

$$P \equiv \mu Q^2$$

and because μ is much higher at start-up than during a regular run, to avoid pressure shock, the feed Q has to be kept low, according to the material, and from the die size, to 20–50% of the output. Thus, a well-preheated barrel to temperature slightly higher than the extrusion temperatures, an rpm approximately 20% lower, and feed much lower than the regular extrusion rate will give the machine a smooth start. Only when the operator is fully conversant with this type of machine can the start-up procedure be shortened and start-up at close to normal rpm and feed be achieved without problems.

As soon as the first material exits the die, temperature and rpm can be slowly brought up to production speed, and feed can be increased. Any increase of feed, however, has to be made in very gradual steps.

In these machines, an increase of screw speed produces its effect in three stages. First the machine extrudes a material of viscosity μ and an average viscosity $\bar{\mu}$ in the screws, therefore having a certain number N of flights filled and using a certain amount of power. After increasing n, the power consumption increases by the square of n as momentarily both $\bar{\mu}$ and N remain the same. Then $\bar{\mu}$ decreases due to the increased localized shear, and the number of filled turns increases.

$$N\uparrow = f\left(\frac{\mu^2}{\bar{\mu}\downarrow}\right) \quad \text{and} \quad Z_t\uparrow = f\,(\bar{\mu}\downarrow; N\uparrow; n^2\uparrow)$$

The power consumption at this stage may show a little increase. Finally, also, the viscosity at the die μ decreases, and so does the pressure needed for extrusion as:

$$P\downarrow = \frac{Q\mu\downarrow}{K_f} = f\,(\mu\downarrow)$$

and the total number of flights $N = f\,(\mu^2)$ decreases rapidly so that:

$$Z_t = f\,(\mu\downarrow; N\downarrow; n^2\uparrow)$$

and the power used settles at or very slightly higher than the previous mark, almost equal to the one before the increase of rpm.

An increase in the amount of material fed affects the power consumption in two stages, and it is sharply felt by the extruder. By increasing the feed, we increase the filling ratio of the screws, and when more material is within the screws, as $N = f(Q)$, there is an increase in the number of filled flights. More power is required for shearing and pumping, and we will have, then, at first a sharp increase of Z. Indeed, keeping all other parameters constant, the values of Q and Z would be directly correlated.

Thereafter, as N is also a function of $\mu^2/\bar{\mu}$, as soon as the effect of the shear is felt and $\bar{\mu}$ decreases, N is also reduced, and so is, although slightly, the power required.

We have, then, as a first step:

$$Z\uparrow = f\,(Q\uparrow; N\uparrow) \quad \text{and then:} \quad Z\downarrow = f\,(\bar{\mu}\downarrow)$$

When less viscous material reaches the die, μ also decreases, and because, as we have seen, N is a function of $\mu^2/\bar{\mu}$, the number of filled flights decreases greatly, and so does the power used:

$$Z = f(\bar{\mu}\downarrow; N\downarrow)$$

The net result is then a sudden large increase in power requirement followed by a slight reduction and, after the feed increase reaches the die, a further slight reduction of Z.

The relationship between Q and Z is then in reality not linear and a 20% feed increase, after a marked temporary increase, the amount of which depends on how carefully the feed increase has been made, may mean only a 10% final increase in power consumption.

To see the final effect, however, enough time has to elapse in order for the material to exit the die with its new lower viscosity and at the newly established rate. Feed increase therefore has to be made in small increments and at somewhat lengthy intervals.

This reaction of the extruder to an increase in feed Q is valid, however, only when the filling ratio Q/Q_c is not close to 1 or when N is less than the available number of flights in the pumping section, or in other words, when the extruder is not close to capacity. Otherwise, other and much more complicated phenomena have to be taken into account, and any feed increase requires more care and time.

A careful watching of the motor ampmeter can give the operator a good idea of what is going on within the extruder.

An increase in rpm will show immediately as an increase in amps, followed after a certain time by a reduction to a level slightly higher than the previous. Conversely, a reduction in rpm will immediately but temporarily decrease the amps, followed, after a certain time, by a marked increase that will level off and, only after more time has elapsed, reduce slowly to a lower level.

An increase of feed will not be shown by the ampmeter until the material reaches the first pressure zone, for example, a counter-flighted section, but at that point in time, a marked increase in amps will be noticeable. Slowly, after a certain time, this value will decrease from its peak and return to lower values, although definitely higher than previously.

Reducing the feed will relieve the extruder very rapidly, but after a certain time and after the extruder stabilizes, the amps will again increase, though to a value lower than at the start of the maneuver.

EFFICIENCY

Extruders are thermodynamic machines. Since single-screw extruders must develop frictional heat to function, designers tried to make a screw that would develop the exact amount of heat required to process the material so that the motor would supply the exact amount of energy to heat, melt, and compress the material. They designed extruders, to the extent possible, as thermodynamic machines operating adiabatically, with constant entropy. As every material has a different specific heat, an adiabatic machine would work ideally only for a specified material and at one particular running condition, provided the extruder could be absolutely thermically insulated from the ambient. However, to achieve enough pressure and enough production, the screw speed has to be relatively high and its diameter large; most of the time, the frictional heat produced by the screw will increase the melt temperature too greatly. Cooling systems therefore had to be devised to transfer heat out of the melt. The machine performs no longer an adiabatic transformation where the enthalpy of the material should be $U_2 - U_1 = AL$, but performs a polytropic transformation where $U_2 - U_1 = AL - H$, in which U_2 is the final-state enthalpy of the material, U_1 is the initial-state enthalpy, A is the thermomechanical equivalence factor, L is the mechanical work, and H is the heat quantity.

Twin-screw extruders are designed as polytropic machines in which the mechanical work does not supply the total energy, but only part of it, and where additional heat is supplied by heaters.

The polytropic transformation performed by this machine can be written $U_2 - U_1 = AL + H$, where the plus sign means that more heat has to be supplied and not removed.

Introducing the insulation value, χ, defined by Jacobi as:

$$\chi = \frac{ln\, f}{ln\, f_{ad}} \quad \text{or} \quad \chi = \frac{\Delta t}{\Delta t_{ad}}$$

where Δt is the temperature differential of the material between entrance and exit, Δt_{ad} is the temperature differential induced in the material by the screw(s) adiabatically, we note that if the temperature increase of the stock is totally produced by the screw(s) $\Delta t = \Delta t_{ad}$ and therefore $\chi = 1$ (adiabatic).

Single-screw extruders or extruders in which the heat to increase the stock temperature is totally produced by the screw but in which some heat has been removed have $\Delta t < \Delta t_{ad}$ and therefore $0 < \chi < 1$ (autogenous). Twin-screw extruders on the contrary, fall into the category in which $\Delta t > \Delta t_{ad}$ or $\chi > 1$ (polytropic).

The big difference between extruders with an insulation value $\chi < 1$ and those with $\chi > 1$ lies in the fact that in the first case the final stock temperature is less than the one induced in the material by the mechanical work of the screw, whereas in the second case, the final temperature is greater than that.

As can be seen in Figure 6-1, the case in which $\chi = 1$ is ideal but only theoretical because there are always some heat losses. When $0 < \chi < 1$ (single-screw extruders) sometimes Δt_{ad} is so big that the material can reach the decomposition temperature, but even if this can be avoided, some energy always has to be spent to decrease the stock temperature by $-\Delta t$.

When Δt_{ad} (the increase in temperature due to the frictional heat) is large, and/or Δt (which represents the extrusion temperature at the die) has to be kept low, a great amount of cooling is required. On the other hand, the $-\Delta t$ (which is the barrel cooling) cannot exceed certain limits because if the cooling is too great, a layer of material will "freeze" near the barrel's inside surface. This will increase the frictional work of the screw, and Δt_{ad} will increase at an even higher rate. The result is that in an autogenous extruder (with

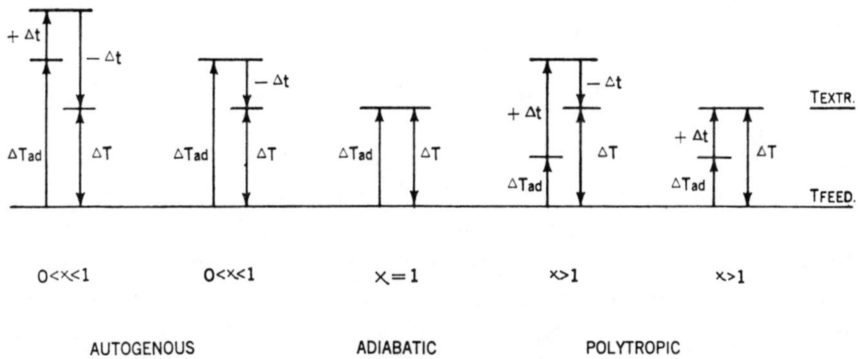

Fig. 6-1. Diagram showing how extrusion temperature is reached in autogenous, adiabatic and polytropic extruders. In autogenous extruders, the mechanical work brings the stock to a temperature higher than extrusion temperature.

insulation value $0<X<1$) the material that reaches a certain temperature by mechanical work cannot be cooled in excess of a specified limited amount.

When $X>1$ (twin-screw extruders) the frictional heat Δt_{ad} is slight, and a certain amount of heat always has to be added. If, after heating, cooling should be needed, that can be easily accomplished because even if the frictional work increases, Δt_{ad} will always be less than the Δt required for extrusion. This allows the extrusion of materials like polystyrene or polyethylene foam, which has to be melted first and then extruded at lower temperature.

In an extruder, as in any machine, energy output is always equal to energy input regardless of the form into which it might be transformed. The energy may be furnished by two sources, mechanical and thermal. The mechanical energy is supplied by the drive motor; the other type of energy, thermal, is supplied positively or negatively by electrical heater bands or by cooling devices around the barrel.

In twin-screw machines, where the two screws rotating in the same direction at low speed do not produce an excessive amount of frictional heat, there is need to add some heat (+ H); in a single-screw extruder, there is need to eliminate heat; therefore, a cooling device is provided (– H). In twin-screw extruders, the total combined energy supplied to the machine is used almost completely for processing the material to be extruded with only a minimum amount not utilized in processing.

From the previous statements, the following equation can be written:

Mechanical energy + thermal energy = utilized energy + lost energy

The mechanical energy (E_m) is used partly to rotate the screws and partly to process the material in the barrel. The portion available for processing of the material is the total mechanical energy multiplied by the mechanical efficiency of the machine. Taking into account the bearings, the various reducing gears, and the belt drive, this can be calculated at $\eta_m = 0.81$.

The total energy input must not only heat the material from its actual temperature (t_{me}) to a stock temperature (t_{ms}), which energy can be designated E_g, but also compress it at a pressure of around 2,000 or 3,000 psi. Therefore, we must consider a melt and plasticating energy E_g and a pressure energy E_p.

Analyzing the energy lost in processing as heat, it can be seen that it is lost in two ways: by radiation and by convection. The first (E_i) can be calculated with the known Stefan-Boltzmann formula, and the second (E_c) is derived by the Wamsler-Hinlein formula. Since the temperature-control instruments call for heat only at various intervals to maintain the required temperature, the thermal energy may be measured by a wattmeter and will be indicated as E_t. The energy balance formula is then written as follows:

$$E_m \eta_m + E_t = E_g + E_p + E_i + E_c$$

For this study, we have taken an LMP Colombo medium-size co-rotating twin-screw extruder, at fixed production rate, using rigid PVC, and have measured the energy input (mechanical and thermal) and the temperature and pressure. The preceding formula in its complete form is:

$$E_m \eta_m + E_t = C_p Q \Delta t_m + \frac{QP}{\gamma} + CS\,\alpha\,\Delta t_c + 1.02\,\sqrt[4]{\frac{\Delta t_c}{d}} \cdot S \Delta t_c$$

To express each term uniformly in kilowatts, we have introduced the following coefficients: 1.162×10^{-3} for E_g, E_i and E_c; 1/36,700 for E_p. The following measured values were then inserted: room temperature $t_a = 18°C$; cylinder temperature $t_c = 175°C$; material temperature, when fed $t_{me} = 20°C$; stock temperature $t_{ms} = 195°C$; from which Δt_m (material) = 175°C and Δt_c (cylinder) = 157°C; pressure $P = 180$ kg/cm^2; output $Q = 234$ kg/hr; cylinder diameter $d = 0.3\ m$; cylinder surface $S = 2.5\ m^2$. Also, specific gravity = 1.4 kg/dm^3, specific heat $C_p = 0.32$ cal/°C. gm, as well as the coefficients that are given by the tables: $c = 3.7$ kcal/m^2 · hr · °C; and $\alpha = 3$.

By calculating the equation with the measured data and given coefficients, we get:

$$(11.4) \times (0.81) + 14.6 = 15.73 + 0.82 + 5.06 + 2.29.$$

The total energy ($E_t + E_m$), which, measured by wattmeters, amounts to 26 kW, is divided so that approximately 7.3 are lost as heat ($E_i + E_c$), 2.1 kW are used to drive the screws ($E_m - E_m \eta_m$), and 16.6 kW ($E_g + E_p$) are used for processing 234 kg of material. The energy used for melting the material is derived from the mechanical source only to the extent of:

$$E_{mu} = E_m \eta_m - E_p = (11.4)\,(0.81) - 0.82 = 8.41 \text{ kW}$$

and the balance is derived from the thermal source:

$$E_{tu} = E_t - (E_c + E_i) = 14.6 - (2.28 + 5.06) = 7.25 \text{ kW.}$$

In the average, depending on screw configuration, the material extruded, and the die used, we have:

	Drive 26%		Losses 50%	
Motor 100%				Heaters (noninsulated) 100%
	Processing → 74%	Material 59% 41%	← Melt 50%	

and in a larger machine with longer screws and well-insulated barrel, we have on the average:

	Drive 20%		Losses 10%	
Motor 100%				Heaters (insulated) 100%
	Processing → 80%	Material 52% 48%	← Melt 90%	

Other tests have been made where the energy input and the temperature of the melt have been measured.

The barrel was arbitrarily divided into a rear and a front part, and within these zones the temperature of the melt, and therefore its enthalpy increase, was measured. Assuming that a noninsulated barrel loses, as we have seen before, approximately half of the energy, we can summarize the results of these tests in Table 6-1. The total energy input, mechanical and thermal, have been divided by the output of the extruder to obtain the unitary value per pound. The total mechanical energy measured was 25Wh/lb of material extruded, which corresponds to a specific production of 30 lb/hp · hr.

Table 6-1. Showing How Energy from the Motor and Heaters Increase the Material's Enthalpy (and Its Temperature) in a Twin-Screw Extruder.

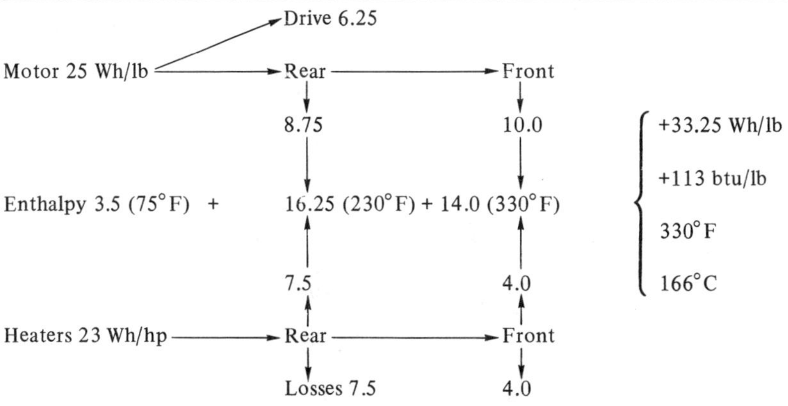

In some cases, when the temperature of the melt within the barrel has to be higher and the final temperature lower as in the case of foam extrusion, there is unavoidably the need for cooling of the material before the material exits the die. In this case, in the front part of the barrel, the material is more viscous, as it is at lower temperature, and the screws at this point put more power than usual into it. Furthermore, no heat will be added to the front part; instead, cooling devices will subtract some. The tests, made in a larger extruder with cooling section at the end of the barrel, give the average the results shown in Table 6-2.

Calculating the energy content for different thermoplastic materials, we can see that, depending on the type of material and its extrusion temperature, it is in the range of 0.03–0.15 kWh/lb. In twin-screw extruders, only part of these 30–150 Wh/lb comes from the motor, as only part of the needed energy is furnished by the motor, and the heater bands supply the rest.

In an adiabatic extruder, where all the energy is furnished by the motor through the screw, only a limited amount of material, determined by the horsepower available, can be processed. Deducting approximately 20% of the motor power, which is needed for the drive, the specific production of these machines can be only from 4–19 lb/hr per each horsepower of the main motor installed.

These calculations, confirmed by practical experiences, show that the specific production of production per unit of horsepower is much larger in twin-screw than in single-screw extruders. Based on

Table 6-2. Showing How Energy from the Motor and Heaters Increase and Subsequent Cooling Decreases Material's Enthalpy (and Its Temperature) in a Twin-Screw Extruder for Production of Foam.

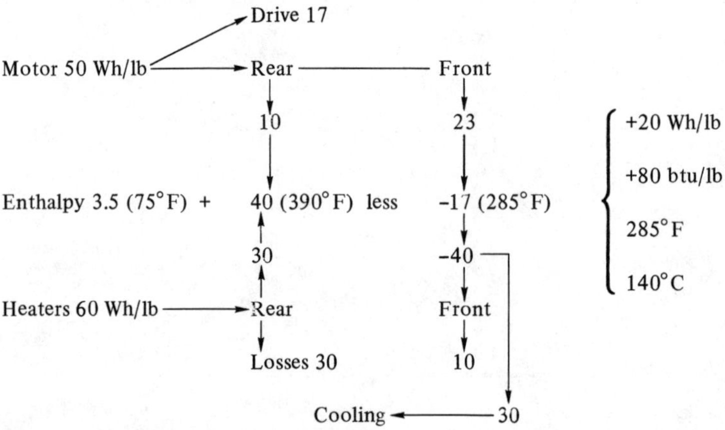

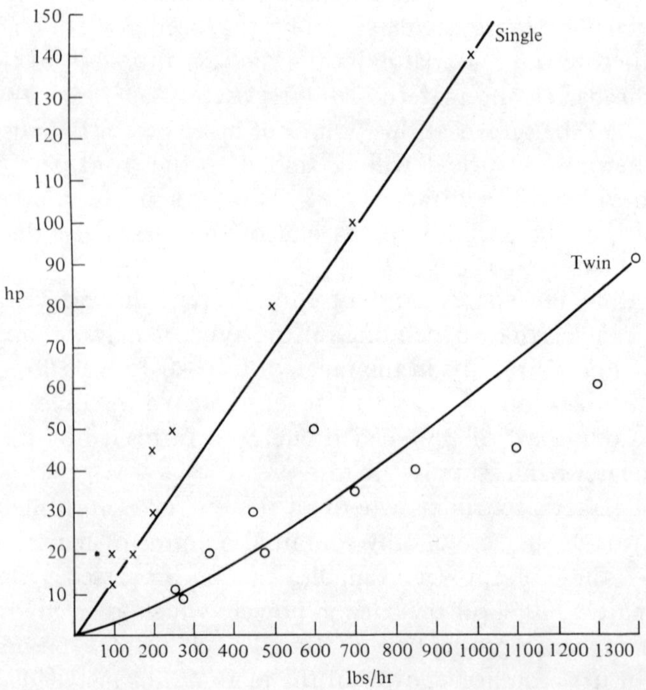

Fig. 6-2. Approximate values of hp motor required with single- and twin-screw extruders (material: rigid PVC).

Table 6-3. Single-Screw Extruders Specific Production with Various Materials, in lb/hp · hr (From A. Griff, *Plastics Extrusion Technology*).

Rigid PVC	7 - 10	Polypropylene	5 - 10
Plasticized PVC	10 - 13	L D P E	7 - 10
Polystyrene	8 - 12	H D P E	4 - 8
A.B.S.	5 - 9	Nylon	8 - 12

published data, single-screw extruders have a production output per rated horsepower that ranges from 4 to 13 lb/hp · hr (Table 6-3).

In corotating twin-screw extruders, depending also on the material used, production averages 25 lb/hp · hr, and as the calculation for the test previously described (p. 111) shows, can easily reach:

$$\frac{Q}{E_m} = 0.745 \ \frac{234}{11.4} = 15 \text{ kg/hp} \cdot \text{hr} = 33 \text{ lb/hp} \cdot \text{hr}$$

For equal output, therefore, twin-screw machines have much smaller motors than corresponding single-screw extruders; for an output of, say, 500 lb/hr of PVC, a main motor of only 25 hp would be sufficient, whereas, for the same output in a single-screw extruder, a main motor of 80 hp would be required (Fig. 6-2).

7
Application of Twin-Screw Extruders

The field of application of twin-screw machines is as wide as the field of extrusion itself. Many polymers are now commercially extruded in single- as well as in twin-screw extruders; these latter, however, due to their positive propulsion and positive mixing, which are characteristics of these machines, can process some of the polymers easier and better. Whereas some material can be easily extruded with single-screw extruders, some material (ultrahigh-molecular-weight polyethylene [UHMWPE]) can only be processed with twin-screw extruders.

Polyolefines are among the materials less sensitive and among the easiest to extrude. Any commercial extruder can give good results with these materials.

For low- and medium-density polyethylene, a temperature range of 135–170°C is usually required, and extrusion of such materials does not present any particular problem.

When the molecular weight of the polyethylene is very high, like UHMWPE with molecular weight over 3.5 million, the situation is very different.

This material, which, when melted, has a very high viscosity, has also the peculiar characteristic that under a certain shear rate its viscosity drops suddenly to a comparatively low value. This characteristic makes extrusion a difficult task for single-screw extruders where extrusion depends on the adhesion of the material to the barrel. The shear developed by the screw flight tips against the barrel renders the material much less viscous at that point, so that it forms a layer of very fluid material that lubricates the barrel surface: the polymer will slip on the barrel's inside surface, thereby hindering and finally stopping the extrusion.

For this material, an extruder in which the pumping is positive and independent from friction against the barrel is absolutely required. Corotating twin-screw extruders, according to Hoechst, the first and one of the few manufacturers of this material, have demonstrated to be the only effective machine for the extrusion of UHMWPE. Even on these machines, however, the above-mentioned characteristic of this material makes the extrusion process very "touchy," and careful attention has to be given to screw speed and temperature profile. Very low rpm and temperature in the range of 180–200°C are satisfactory.

Polypropylene, another resin of the same family, behaves very much like low-density polyethylene but develops less frictional heat, so that the barrel heaters have to be kept at a slightly higher temperature to compensate for it.

All various types of polystyrene also do not present any particular problem in extrusion with any type of extruders. Temperatures may have to be kept a little higher with rubber-modified high-impact grades. One exception is when it is desirable to use polystyrene as it comes out of the polymerization process, that is, in the form of tiny beads or pearls. The use of these beads saves the cost of their pelletizing into small cubes or pellets. Direct extrusion of these beads into finished products cannot easily be performed in single-screw extruders, as it is difficult to control the feed of this rolling material and its regular advancing in the first few flights of the screw. With this material, the separate feed device of most of twin-screw extruders, together with the very large and deep channels of the feed section of the screws and their positive pumping even in the feed zone, are of great advantage.

Another type of material with special characteristics are nylon and other polyamides; these materials, very viscous when still not properly melted, become suddenly very fluid with a slight increase in temperature. A positive pumping machine is of great advantage with these materials, and a not too accurate temperature control is sufficient. High melting temperatures, above 225°C, however, require very strong heater bands with very high specific load, which may affect their durability.

The same high specific load heaters are required with polycarbonates, which also have a high melting temperature.

Both nylons and polycarbonates are easily extruded with twin-screw extruders because of their positive pumping, which, fluid as the first material or viscous as the latter may be, always propel the material forward.

Polyvinyl chloride is used in conjunction with various additives, and the formulation depends mainly on the characteristics required of the final product. The formulation, however, also depends on the type of machine used, as different extruders treat the material differently. Two of the additives especially are in the formulation because they are required in the extrusion process: heat stabilizers and lubricants.

Single-screw machines develop high shear, which adversely affects the polymer, and some decomposition occurs. While stabilizers really only prevent discoloration, the lubricants facilitate extrusion and thereby limit decomposition. Lubricants, however, cannot be added over a certain quantity in single-screw extruders; otherwise, friction of the material against the barrel is lost, and the machine ceases to extrude; lubricants can be added in much larger amounts in twin-screw extruders, as these, with their positive pumping action, always extrude.

Among the most used lubricants are waxes, polyethylene waxes, dibasic lead stearates, and soaps of other metals like cadmium, barium, etc., glycerols, and similar products.

PVC compounds, formulated for use with single-screw extruders, do not always work well with twin-screw extruders, and usually it is better to formulate the compound especially for these latter. The more gentle action of the deep channels and low rpm of twin-screw machines on the PVC, together with low residence time, allow the compound to be formulated with much less stabilizer, a costly ingredient, and with more lubricants.

Furthermore, the high mixing capacity of these machines can be utilized for extruding PVC, blended but not yet pelletized. The use of dry blend not only saves the expenses of the pelletizing operation, but as the material undergoes one and not two extrusion processes, even less stabilization is required. Rigid PVC without a drop of plasticizer can be extruded on twin-screw extruders. Typical lead base formulation for a rigid PVC product (pipes) to be extruded directly from dry blend through a twin-screw extruder may be:

PVC	100 parts
Stabilizer	2.5–3 parts/100 resin
Lubricant	3–3.5 parts/100 resin

With this dry blended formulation, temperatures must be kept in the 165–180°C range.

Soft PVC is obtained with the addition of plasticizer. The most common among them are Dioctyl phthalate, Dibutyl phthalate, Dioctyl hexyl adipate, Tricresyl phosphate, and some polymeric products.

Plasticizer increases the flexibility of PVC at room temperature, and some of them, in appropriate quantity, keep PVC flexible even at temperatures below 0°C.

Plasticizers also render the melt consistency similar to that of a rubbery mass, so that, in most cases, plasticized PVC creates no problem in extrusion either in single- or twin-screw machines.

A typical example of soft PVC formulation for twin-screw extruders may be:

PVC	100 parts
Plasticizer	60 parts/100 resin
Stabilizer	2 parts/100 resin
Lubricant	2.5 parts/100 resin

Extrusion temperature in twin-screw extruders are much lower, in the range of 140–160°C, and there is less danger of decomposition. The self-cleaning characteristic of twin-screw machines is of great advantage in the extrusion of PVC, whether rigid or flexible, but, most important, when rigid, because it prevents particles of the material being left behind, stagnant and decomposed.

SENSITIVE MATERIALS

In order to melt, any material to be extruded has to increase its temperature; according to its melting point and its specific heat, a certain number of BTUs have to be furnished to it, and we should never forget that in twin-screw extruders half of those BTUs must come from the barrel heaters.

Extrusion conditions are affected not only by the characteristics of the resin and the type and amount of additives but also by the type of machine and its screw configuration. Furthermore, depending on the residence time of the polymer up to the melting zone, a certain barrel temperature is needed: different barrel temperatures are therefore required for different materials. (See section on Melting.)

If too little heat is furnished, the material may not melt properly, and twin-screw machines, trying to extrude it anyhow because of their positive pumping, may overload; if the heat is too great, the material may degrade or decompose, as the case may be.

Degradation of polymers occurs with any machine every time the material is extruded, but if proper temperatures are maintained, degradation is much less in twin- than in single-screw extruders. This is because in single-screw extruders as well as in some counterrotating twin-screw extruders, there are points at which high shear is developed. This localized high shear cuts the lengthy molecules of the polymer and reduces its viscosity. With this, also the mechanical properties of the finished product are lowered.

Every time a polymer is extruded, there is a slight reduction of its solution viscosity, but as the following table shows, this reduction is much lower if a twin-screw extruder has been used for the extrusion of that material (Table 7-1).

Decomposition is when the polymer's molecule is affected and atom substitution or liberation occurs. This is the case of polyvinyl chloride (PVC) which is thermally unstable, that is, if exposed to heat for a period of time, it slowly changes its structure (CH_2=CH–C1) liberating chlorine.

Against this phenomenon, heat stabilizers are added to PVC in the formulation before extrusion. These usually consist of metallic compounds, lead carbonate, complexes of barium-cadmium, or some products like organotins (methyl-, octyltins) and mercaptanes, which all can be incorporated in PVC formulations.

Stabilizers, however, do not really avoid the process of decomposition but only avoid the dark-brown coloration that is the effect of this process.

With sufficient heat and/or shear for a sufficient time during extrusion, all PVC composition will more or less decompose. Little

Table 7-1. Decrease of the Solution Viscosity after Extrusion (Polystyrene).

	After One Extrusion Through:	
Virgin Material Solution Viscosity	Twin Corotating Screw Extruder	Conventional Single-screw Extruder
30	25	23
26	25	22
20	19	17

shear, controlled heating, and a short barrel, which are basic points found in twin-screw extruders, are essential for limiting decomposition to a minimal, tolerable extent even when a very little amount of stabilizers are added to the PVC.

FOAMS

Extruding thermoplastic foams, especially of polystyrene, polyethylene, and PVC represent another good use to which twin-screw extruders can be put.

The method of direct injecting a gaseous blowing agent midbarrel during extrusion has proved more economical than chemical blowing agents, which decompose liberating gas into the melted mass.

The injection of gas under pressure, however, requires that before the point of injection, the extruder is able to create enough pressure so as to counteract the pressure of the gas, thus preventing it from flowing backward along the barrel and escaping through the feed opening. In single-screw extruders, this requires a long metering zone before the injection port, even with the presence of a special sealing device in the screw (blister) and therefore a long barrel (up to 50/1 L/D). Intermeshing twin-screw extruders, with their positive pumping action, can easily provide the necessary pressure, and the addition of devices like a counterflighted section assures a tight seal under any condition. More important, in order to dissolve the gas into the resin completely and well, the melt has to be maintained at a temperature that is higher than the required extrusion temperature. On the other hand, to obtain good expansion and avoid cell collapsing, the extrusion temperature must be slightly above the freezing temperature of the material; a good cooling is then necessary.

In single-screw extruders, the heat developed by shear is very high; but in these machines, cooling cannot be very intense, otherwise the material near the barrel's inside surface would freeze, and the frictional heat would increase, thus defeating the purpose of the cooling of the material.

To circumvent this problem when single-screw extruders are used, the gas injection and the mixing at high temperature are made in one extruder, and the material is then transferred into a second machine specially designed and rotating at very low rpm, which does the cooling: two extruders in tandem are necessary. In some cases the addition of some kind of static cooler-mixer helps the cooling operation. The sectional construction of the screws of some of the intermeshing twin-screw machines and their very low rpm allow the whole operation to be done in one extruder only.

As we have seen, these machines can develop very little heat through mechanical work, far below the amount required for the material to reach the temperature required for gas mixing; additional heat is applied in order to reach it. After the mixing, no more heat is applied, and due to the low shear developed, the temperature of the melt will decrease; because of the relatively slow movement down the barrel, with a reasonable little cooling its temperature will easily reach the one required for extrusion. One extruder can therefore perform the whole operation. Polyethylene sheet and tubing, polystyrene sheets and slabs and PVC profiles foamed to densities in the order of 3 to 2 pcf. have been successfully extruded with twin-screw extruders.

An example of the energy balance in one such operation with a twin-screw extruder is given in Table 6-2. Simplicity of operation, saving in investment – one extruder instead of two – and the day-to-day energy saving make twin-screw extruders outstanding for foam extrusion.

CONCLUSION

The very special principle of operation that characterizes corotating intermeshing twin-screw extruders can be summarized as follows:

- The screws can be starve-fed by an independent mechanism so that screw speed setting does not affect output.
- Positive forwarding action eliminates problems like the bridging or slipping of solid material in the feed section, allowing the feed of pellets, beads or powder.
- Shear rate is very low, and additional external heat, much easier to control, is added and shear stresses distribution can be adjusted by changing screw speed and/or output.
- Stock temperature can be raised to the desired level without exceeding it, and low temperature extrusion is easily obtainable.
- Positive pumping in melted state eliminates surging and promotes better dimensional control of the extrudate.

- Mixing is also positive and accomplished without overheating the material.
- Better dispersing and homogenization.
- Self-cleaning is fast and complete and accomplished without overheating the material.
- Lower HP required per rated output, resulting in lesser stresses on the machine.
- Energy saving as the required energy is very close to the theoretical value needed.

In a twin-screw extruder where a great part of the needed heat is given through the barrel heaters, an increase in screw rpm increases the shear, which is very small, but also reduces the residence time of the material in the barrel. Depending on the running condition, a change in screw rpm will produce only slight variations in the stock temperature, which can easily be readjusted by decreasing or increasing the barrel heater temperature. If, for example, too much shear is evident in a single-screw machine during extrusion, screw speed has to be reduced. The result is a proportional decrease in production. In a twin-screw extruder, a reduction of the screw speed reduces the flow capacity, but the same feed is maintained and therefore the actual production remains constant.

With the same die, to increase the head pressure in a single-screw extruder, the screw speed has to be increased which results in high shear. As the propulsion as well as the pressure build-up depend on friction and viscosity, these influence in both cases the stock temperature and output.

In a twin-screw extruder, the screw's speed does not affect pressure, nor does it affect the propulsive action, so that higher head pressure can be obtained with the same screw speed by greater feeding. For processing materials with a high viscosity to shear ratio, and low specific heat, these differences become particularly evident.

Therefore, these extruders are especially suitable for the extrusion of heat-sensitive polymers, for products that have to be extruded at low temperature, for the extrusion of uncompounded dry blends, and for materials that present difficulty in feeding or that have to be thoroughly degasified.

The advantage of being able to combine different sections so as to achieve a screw configuration that performs on the material the work required at the point where it is required gives these extruders the capability of easily handling different tasks on different materials.

Due to their always present positive displacement, there are no problems, as we have already mentioned, in feeding slippery or rolling materials as highly lubricated compounds or polystyrene small beads as they come out of the reactors or to feed them with melted low-viscosity materials.

The low rpm of the screws together with a low shear configuration handles the material delicately enough to allow good extrusion at relatively low temperature of very lightly stabilized rigid PVC compounds or dry blends. The resulting product, with low ash residual and minimal lowering of the K value, shows excellent mechanical properties.

ABS powder compounding, extruded with screws having one venting between two mixing zones, results in very homogeneous and nonporous pellets.

Parison for blow molding, extruded at temperatures of 190–200°C for PVC, and 140–160°C for PVC or polypropylene copolymers, can be made with screws with the length of the mixing zone reduced to a minimum.

Short screws, simply composed of a short feed and slightly longer pumping zones, have proved good for extrusion, or better, for pumping, of liquid polyurethane.

Polyesters and polyamides can easily be degased with screws having two venting sections, one of which is under high vacuum.

The self-cleaning action of the screws avoids hangups, and the low residence time avoids specks of degraded material in the extrusion of the very heat sensitive PVC with only organic stabilizers and of polyvinylidene chloride.

Polyolefine resins, especially UHMWPE, with its inherent lubricity and its abrupt lowering of viscosity under shear, can be extruded with screws almost totally composed of pumping sections.

Finally, with the proper screw configuration, foamed products either in polyethylene or polystyrene can be produced by mixing additives, melting at relatively high temperature, injecting the blowing agent, and then cooling the melt to the lower temperature required for foaming, performing all these operations in one extruder.

Appendix

MANUFACTURERS AND DISTRIBUTORS OF TWIN-SCREW EXTRUDERS IN THE UNITED STATES

Corotating Extruders

Baker Perkins Inc.	1000 Hess St. Saginaw, MI 48601 (517) 752-4121
Berstorff Corporation	P.O. Box 240357 Charlotte, NC 28244 (704) 523-2614
Gerd Lester Corporation (LMP Type)	60 East 42nd St. New York, NY 10017 (212) 867-0789
Haake Inc. (Baker Perkins – W & P Types)	224 Saddle River Rd. Saddle River, NJ 07662 (201) 843-2320
Japan Steel Work Amer. Inc. (Krauss Maffei Type)	200 Park Ave. New York, NY 10017 (212) 867-5600
Toshiba Machine Co. of America	7964 Kentucky Dr. Florence, KY 41042 (606) 525-1616
Werner & Pfleiderer Corp.	663 East Crescent Ave. Ramsey, NJ 07446 (201) 327-6300

Counterrotating Extruders

Amacoil Machinery	111 Plain Ave. New Rochelle, NY 10801 (914) 235-5050

American Leistritz Corp.	369 S. Miguel Dr. Newport Beach, CA 92660 (714) 640-1083
Brabender Instruments Inc.	50 E. Weseley St. S. Hackensack, NJ 07606 (201) 343-8425
Cincinnati Milacron Inc. (Anger, AGM Types)	4165 Halfacre Rd. Batavia, OH 45103 (513) 536-2000
Egan Machinery Co.	36 S. Adamsville Rd. Somerville, NJ 08876 (201) 722-8000
Haake Inc. (Anger Type – Laboratory equipment only)	224 Saddle River Rd. Saddle River, NJ 07662 (201) 843-2320
Hopkin Hunt Co. Inc.	P.O. Box "A" Hampton, NH 03842 (603) 926-3058
Japan Steel Work Am. Inc. (Krauss Maffei Type)	200 Park Ave. New York, NY 10017 (212) 867-5600
Krauss Maffei Plast. Mach. Div.	P.O. Box 9104 Wichita, KA 67277 (316) 945-5251
Maillefer Co.	749 New Ludlow Rd. S. Hadley, MA 01075 (413) 534-3306
Reifenhauser-Nabco (Schloemann Tipe)	P.O. Box 590 Springfield, VT 05156 (802) 886-8321
N R M Corporation	180 South Ave. Tallmadge, OH 44278 (216) 663-1600
Toshiba Machine Co. of America	7964 Kentucky Dr. Florence, KY 41042 (606) 525-1616

FOREIGN MANUFACTURERS OF TWIN-SCREW EXTRUDERS

Corotating Extruders

Berstoff Maschinenbau	Postfach 629 Hannover, West Germany
Bone Bros Ltd. (Egan Type)	Manor Farm Road Alperton, Middlesex, England
Creusot-Loire (LMP Type)	Usine de l'Ondaine Firminy, (Loire), France
Ikegai Iron Works (LMP Type)	2 Mits-Shikokumaki Shiba-Minato Ku Tokyo, Japan
Japan Steel Works (Krauss Maffei)	12 Banchi-1 Chome, Yuracucho Chiyoda-Ku Tokyo, Japan
LMP, Lavorazione Materie Plastiche	Casella Postale 305 10100 Torino, Italy
Novelle Mapré S.A.	Diekirch, Grand Duche de Luxembourg
Werner & Pfleiderer Masch. Fabr.	Theodorstrasse 10 Stuttgart-feuerbach, West Germany

Counterrotating Extruders

AGM Gmbh (Anton Anger)	Huehnersteig 9 Linz, Austria
W. Anger OHG	Bahnhofstrasse 28 Munich-Unterforing, W. Germany
C. Bausano	Corso, Indipendenza 111 Rivarolo Canavese (Torino) Italy
Japan Steel Work (Krauss Maffei Type)	12 Banchi-1 Chome, Yuracucho Chiyoda-Ku Tokyo, Japan
G. Kestermann KG	D 4970-Bad Oeynhausen, West Germany

P. Leistritz GmbH	Nurenberg, West Germany
Moi SpA	Via dei lavoratori 78, Cinisello Balsamo (Milano) Italy
Reifenhauser KG	Spicher Strasse, D 5210-Troisdorf, West Germany
Schloemann AG	Postfach 7240 Duesseldorf, West Germany

Bibliography

1. AGM Cotruder with conical screws. *Plastics Technology,* April 1966; *Intern. Plast. Engineering,* March 1966; *Plastics,* September 1965; *Kunstoff und Gummi,* June 1965.
2. Anger, Standard twin screw machines. *Int. Plast. Eng.,* March 1962; *Industrie des Plast. Modernes,* December 1961.
3. Bayer, S. Der Colombo Extruder. *Kunstoffe,* April 1968.
4. Benadi, A. Il moto delle resine termoplastiche nell'interno delle viti degli estrusori. *Poliplasti,* July 1967.
5. Banadi, A. Geometria delle coclee di estrusione e scelta di progetto per le nuove applicazioni degli estrusori bivite. *Congresso Internazionale delle Materie Plastiche ed Elastomeriche,* Milano, Italy, October 1968.
6. Bernhardt, E. *Processing of Thermoplastic Materials.* New York, USA, Van Nostrand, 1958.
7. Bone Bros. Twin screw machines. *Plastics.* June 1965.
8. Chung, C. New ideas about solid conveying in screw extruders. *SPE Journal,* May 1970.
9. Colombo, R. Tecnologia di estrusione delle resine espanse. *Materie Plast. ed Elast,* August 1969.
10. Darnel, W. -Mol, E. Solid conveying in extruders. *SPE Journ,* April 1956.
11. Derex, Twin screw extruders. *Int. Plast. Eng,* August 1963.
12. Doboczky, Theoretishes und wirklishes Austossung der doppelshnecken extruder. *Plast. Verabeiter,* June 1965.
13. Fisher, E.G. Extrusion of Plastics. London, England, Iliffe, 1964.
14. Gregory, R. Friction coefficient of plastic and steel. Chicago SPE Antec, 1969.
15. Griff, A. Plastics Extrusion Technology. New York, Van Nostrand, 1968.
16. Herrmann, H. Mixing and self-cleaning characteristics of single screw and twin screw extruders. *Chem. Eng. Tech,* January 1966.
17. Ikegai. Screw die system. *Jap. Plast. Age News,* June 1963.
18. Jacobi, H. Screw extrusion of plastics. London, England, Iliffe, 1963.
19. Jeremko, D. Twin screw extruders; are really a factor in direct extrusion? *Mod. Plastics,* November 1969.

20. Leonard, Italian both twin and single screw extruders. *Int. Plast. Eng.*, June 1965.
21. LMP/Colombo. LMP/Colombo for PVC powder feed – *Kunstoffe und Gummi,* September 1963 – *Int. Plast. Eng.*, June 1962 – Kunstoffe, April 1958.
22. Maekelt, H. Observation on machine construction in the development of twin screw extruders. *Int. Plast. Eng.*, – January 1964.
23. Mapre. Twin screw machines. *Int. Plast. Eng.*, June 1965 – *Kunstoffe,* March 1962.
24. Martelli, F. How to extrude big PVC pipes. *Modern Plast,* September 1965.
25. ——. Twin screw extrusion of expanded PS sheet. *Modern Plast.*, September 1969.
26. ——. Extrusion of Ultra High Molecular Weight Polyethylene. Baltimore SPE Retec, October 1970.
27. ——. Twin screw extruders, a separate breed. *SPE Journal,* January 1971.
28. ——. Extruded foam sheet orientation. San Francisco SPE Antec, 1974.
29. ——. Caluclating extruder energy efficiency. *Plast. Compounding,* April 1980.
30. Menges, G.–Klenk, P. Verarbeitung von PVC hart auf einem Zweischnecken Extruder (Schloemann) *Plastverarbeiter,* December 1966.
31. Moi. Twin screw machines. *Int. Plast. Eng,* December 1962.
32. Prause, J. Twin screw extruders, when to use and why – *Plast. Technology,* November 1967.
33. Prause, J. Twin screw extruders, an engineering analysis – *Plast. Techn,* February 1968.
34. Prause, J. Twin screw extruders, who is who among suppliers – *Plast. Techn,* March 1968.
35. Schenkel, G. Schneckenpresse fuer Kunstoffe – Munchen, W. Germany, Hauser 1960.
36. Schloemann. Twin screw extruders Pasquetti system – *Kunst. und Gummi* August 1965 – *Int. Plast. Eng.* March 1965 – *Kunst. und Gummi* November 1963 – *Int. Plast. Eng.* May 1963.
37. Shooter – Thomas. Frictional properties of some plastics – *Research,* November 1949.
38. Selbach, Neue schneckenform fuer doppelschnecken Extruder – (Kestermann) *Plastverarbeiter,* November 1962.
39. Squires, P. Processing of thermoplastic materials – New York, Reinhold, 1959.
40. Squires, P. Screw extruder pumping efficiency – *SPE Journ.*, May 1958.
41. Street, L. Twin screw extruders versus single screw extruders – SPE Antec, Chicago 1969.
42. Storch – Chemica. Production of expanded PS Sheet – *Review for the plast. Industry*, September 1969.
43. Todd – Irving. Axial mixing in self wiping reactors – *Chemical Eng. Progress,* September 1969.
44. Todd D. – Energy control in twin screw extruders – Baker Perkins study.
45. Todd, D. – Mixing in starved twin screw extruders – Baker Perkins study.

46. Uhland, E - Dienst, M. Development and tendencies for corotating twin screw extruders - Atlanta SPE Antec 1975.
47. Werner, H. - Eise, K. An analysis of the conveying characteristics of twin corotating extruders - Washington, D.C. SPE Antec 1978.
48. ——. Neue schneckenform fuer Doppelschnecken Extruder - *Plastverarbeiter,* November 1965.
49. Werner - Pfleiderer. Twin screw compounders - *Int. Plast. Eng.,* January 1965 - *Kunstoffe und Gummi,* June 1964 - *Plastics,* December 1962.
50. ——. An analysis of twin screw extruder mechanism, *Polymer Processing News* - Vol. 8 #2, 3 - Summer and fall 1977.
51. ——. Plasticizing and mixing principles of twin screw extruders - *Polymer Processing News* - Vol. 9 #2, Fall 1978.
52. Windsor. Twin screw extruders (LMP) *Int. Plast. Eng.,* January 1965.

Index of Most Commonly Used Symbols

C_e	equivalent circumference
C_p	specific heat
D	screw diameter
D_i	screw root diameter
D'	primitive screw diameter
D_e	equivalent diameter
E	flight tip max width
e	flight tip width
F	frictional factor
h	channel depth
h'	primitive channel depth
I	interaxis
K	conductance of passages
N	number of filled flights
n	screw speed
n_e	speed of equivalent screw
p	pitch
Q	output
Q_c	output capacity
q	back flow
S	channel length
S_e	equivalent channel length
T	residence time
V	volume of material
Z	power
α	flank angle
β	semiangle of intermeshing
$\dot{\gamma}$	shear rate
ϵ	clearance between tip and bottom
η	efficiency
θ	sliding angle
μ	viscosity
$\bar{\mu}$	average viscosity
ν	relative speed of surfaces
ρ	clearance between screw and barrel
σ	clearance between flanks
ϕ	pitch angle
$\equiv$	proportional to
$\uparrow$	increases
$\downarrow$	decreases

Subject Index